Strive for a 5

Preparing for the AP® Human Geography Examination

to accompany

Human Geography for the AP® Course
by Barbara S. Hildebrant, Seth Dixon, Max Lu, Kenneth H. Keller, and Roderick P. Neumann
Second Edition

AF897625

Max Lu
Laura Kmetz
Kolleen Madeck
Greg Sherwin

bedford, freeman & worth
publishers

Strive for a 5: Preparing for the AP® Human Geography Examination
By Max Lu, Laura Kmetz, Kolleen Madeck, and Greg Sherwin
to accompany
Human Geography for the AP® Course, Second Edition

ISBN: 978-1-319-61413-3

Printed in the United States of America.

1 2 3 4 5 6 30 29 28 27 26 25

Bedford, Freeman & Worth Publishers
120 Broadway, New York, NY 10271

bfwpub.com/aphg2e

Table of Contents

PREFACE

This study guide, *Strive for a 5: Preparing for the AP® Human Geography Examination*, is designed for use with *Human Geography for the AP® Course*, Second Edition, by Hildebrant et al. Page numbers for figures called out in the *Strive* guide refer to the textbook. This guide is intended to help you assess your knowledge of the material covered in the textbook, to practice the 5 AP® skills and task verbs, and prepare for the AP® Human Geography Exam. The guide is organized into two sections: a study guide that reviews each unit in the textbook, and a test preparation section.

STUDY GUIDE

You may use this study guide to evaluate your understanding of the content in each unit as it is covered in class or as a review prior to the AP® exam. Each unit contains the following features:

- **Unit openers:** Before you read each unit, read the summary of the unit and answer the "What do you see?" questions about the unit-opening image in the textbook. You will return to that image at the end of the unit.
- **Before you Read:** Review the learning goals and key terms in the module.
- **While you Read:** This section contains questions that encourage active reading of the material in the textbook. Answering these questions will improve your understanding and ability to remember the material.
- **After you Read:** Review the key terms in the module using the matching activity.
- **Unit Review Exercises:** Use these AP®-style practice questions to assess your understanding of the unit and practice your AP® skills.

After you complete each module and unit, reflect on which topics or skills you struggled with and which you excelled at. Make a note for yourself as to which topics or skills you need to brush up on before the AP® exam. Please ask your teacher for all answers to these questions.

THE TEST PREPARATION SECTION

The test preparation section includes two full-length practice exams. The tests each consist of 60 multiple-choice questions and 3 free-response questions, just like the AP® Exam. After completing the exams, ask your teacher for the answer keys to check how you scored. Take note of the questions you answered incorrectly and be sure to brush up on those topics or skills.

Best of luck on the AP® Human Geography Exam!

Unit 1
Thinking Geographically

Unit Summary

The study of geography has been around since the days of the ancient Greeks. To start thinking geographically, one must understand the essential questions of the discipline: "where" and "why". Human geographers are interested in where different human activities take place and why they locate where they do. They are obviously interested in where human activities do not occur and why they do not occur in specific places. In this unit, you will learn about the "geographers toolbox" or the tools geographers use to think geographically. This will be a foundation for the rest of the units.

Modules in Unit 1

Module 1: Introduction to Maps
Module 2: Geographic Data
Module 3: Spatial Concepts and Types of Diffusion
Module 4: Human-Environmental Interaction
Module 5: Scales of Analysis
Module 6: Regional Analysis

What do you see?

Look closely at the unit-opening photograph. Write down 5 things you observe about it. You will return to this image in the AP® Seeing Geography feature at the end of the unit.

1)

2)

3)

4)

5)

Module 1
Introduction to Maps

Before You Read the Module

Focus on Learning Goals

Use the Module learning goals as a way to divide or "chunk" the modules into parts. After you read a section of the module and answer the questions in this book, you should stop and ask yourself: "do I sufficiently understand the goal?"

LG 1-1: Define a map.
LG 1-2: Explain how maps aggregate data and illustrate spatial patterns.
LG 1-3: Identify the different parts of a map.
LG 1-4: Identify the two main types of maps.
LG 1-5: Explain the benefits and drawbacks of map projections.
LG 1-6: Describe how to think critically about maps and map projections.

Pre-Read the Key Terms

Take a look at the key terms you will encounter in this module.

map
cartographer
data aggregation
spatial perspective
spatial patterns
time–distance decay
map symbols
legend
compass rose
absolute direction
map scale
scale
absolute distance
relative distance
relative direction
elevation
isoline
topographic map
reference map
thematic map
choropleth map
cartogram
proportional or graduated circle map
dot density or dot distribution map
map projection
Mercator projection
Peters projection
Goode homolosine projection
polar projection
Robinson projection

While You Read the Module

Answer the questions as you read Module 1: Introduction to Maps.

LG 1-1: ***What is a map?***

1) Define map and identify the title of the map in Figure 1.1 (p.14).

2) Define cartographer.

3) Describe the purpose of maps

LG 1-2: ***How do maps aggregate data and illustrate spatial patterns?***

4) Identify a state with a county at the lowest population density in Figure 1.1 (p. 14).

5) Define data aggregation.

6) Describe how the data is aggregated in Figure 1.1 (p. 14).

7) Complete the following table.

Spatial Concepts	*Definition*
Spatial Perspective	
Spatial Patterns	

8) Explain why many world cities are located near bodies of water.

9) Describe the difference between a clustered spatial pattern and a dispersed spatial pattern.

10) Define time-distance decay.

11) Using Figure 1.4 (p. 16), explain why Juan frequently goes to a local bodega but Marisol rarely goes to a large shopping center.

LG 1-3: ***What are the different parts of a map?***

12) Define map symbols.

13) Using Figure 1.6 (p. 17), describe how the arrow symbols help us understand spatial patterns.

14) Define legend.

15) Use Figure 1.7 (p. 17) to complete the following table.

Identify a state with the highest poverty rates.	
Identify a state with the lowest poverty rates.	

16) Define a compass rose.

17) Define absolute direction

18) Define map scale and describe what 1/100,000 or 1:100,000 means on a map.

19) Using Figure 1.9 (p. 19), describe whether a cartographer should "zoom in" or "zoom out" the scale if they wanted to see the spatial pattern throughout California.

20) Elevation is an important concept. Complete the following table.

Define elevation.	
Describe how an isoline map displays elevation.	
Describe how a topographic map displays elevation.	

LG 1-4: ***What are the two main types of maps?***

21) Define reference map.

22) Define a thematic map.

23) Complete the following table.

Type of Thematic Map	*Describe how spatial data is presented*
Choropleth map	
Cartogram	
Proportional or graduated circle map	

Dot density or dot distribution map	

LG 1-5: ***Explain the benefits and drawbacks of map projections.***

24) Define map projection.

25) Complete the following table.

Type of Map Projection	*Describe the strengths and weaknesses of the map projection*
Mercator projection	
Peters projection	
Goode homolosine projection	
polar projection	
Robinson projection	

LG 1-6: ***Describe how to think critically about maps and map projections***

26) Use Figure 1.17 (p. 26) to complete the following table.

Identify the type of map.	
Identify whether each dot represents numbers or percentages.	
Describe how the data is aggregated.	
Describe what this data is telling us.	

Explain how else the data could be presented.	

After You Read the Module

Review the Key Terms Part 1

Match each key term on the left with its definition on the right.

_____ map	1) A drawing, usually found on the edge of a map, showing the four cardinal directions (North, south, east, and west).
_____ cartographer	2) A key to the meaning of the symbols and colors on a map.
_____ data aggregation	3) A two-dimensional representation (flat) of a geographic area or place.
_____ spatial perspective	4) The process of collecting and organizing large amounts of information.
_____ spatial patterns	5) The placement arrangement of objects on earth surface.
_____ time-distance decay	6) A person who makes maps.
_____ legend	7) The idea that near things are more related than distant things.
_____ compass rose	8) A geographic perspective that seeks to identify and explain the users of space.
_____ absolute location	9) Corresponds to the direction on a compass: North, south, east, west, and combinations such as northeast and southwest

Review the Key Terms Part 2

Match each key term on the left with its definition on the right.

____ map scale	1) A measurement of the level of social, cultural, or economic similarity between places.
____ scale	2) Distance above sea level.
____ absolute distance	3) The distance on a map in relation to the distance in actual space
____ relative distance	4) A direction that can be described as position, such as in front of or behind, to the left or to the right.
____ relative direction	5) The territorial extent of an idea or object.
____ elevation	6) The distance that can be measured with a standard unit of length.

Review the Key Terms Part 3

Match each key term on the left with its definition on the right.

____ isoline	1) A thematic map that shows data aggregated for a specific geographic area, often using colors to represent different values.
____ topographic map	2) A map that distorts the geographic shape of an area to show the size of a specific variable.
____ reference map	3) A map that uses symbols or different sizes to represent numerical values

____ thematic map

____ choropleth map

____ cartogram

____ proportional or graduated circle map

____ dot density map

4) On a map, a line that connects or links, different places that share a common or equal value.

5) A map that uses dots to represent objects or counts;The dad can represent one object or several objects.

6) A graphic representation of the three-dimensional configuration of the earth surface.

7) A map that emphasizes the spatial pattern of geographic, statistics or attributes.

8) A map that shows geographic locations on earth surface, such as the locations of cities or oceans.

Review the Key Terms Part 4

Match each key term on the left with its definition on the right.

____ map projection

____ Mercator projection

____ Peters projection

____ Goode homolosine projection

____ polar projection

1) A map projection that looks down at the earth from a perspective of one of the poles.

2) A map projection that shows all the landmasses with their true areas, but distort their shapes.

3) A method for representing the surface of the earth on a two-dimensional surface.

4) A map projection that attempts to create the most visually appealing representation of the Earth by keeping all types of distortion relatively low.

5) A map projection that is useful for navigation; however, landmasses become increasingly distorted the further away they are from the equator.

____ Robinson projection	6) A map projection that avoid shape distortion and the restrictions of a rectangular map by creating interruptions in the map's continuity

Module 2
Geographic Data

Before You Read the Module

Focus on Learning Goals

Use the Module learning goals as a way to divide or "chunk" the modules into parts. After you read a section of the module and answer the questions in this book, you should stop and ask yourself: "do I sufficiently understand the goal?"

LG 2-1: Identify who gathers geographic data and why.
LG 2-2: Explain how geographic data is gathered
LG 2-3: Describe the consequences of decisions made using geographic data.

Key Terms

Take a look at the key terms you will encounter in this module.

census
absolute location
longitude
Global Positioning System
remote Sensing
satellite imagery
fieldwork
latitude
Prime Meridian
Geographic Information System
aerial photography

While You Read the Module

Answer the questions as you read Module 2: Geographic Data.

LG 2-1: ***Who gathers data and why?***

1) Identify FOUR ways individual data gatherers collect information according to the reading.

2) Define *census*.

3) Describe ONE way census data is used by the government.

LG 2-2: ***How is Geographic Data gathered?***

4) Define *fieldwork*.

5) Complete the following table.

Locational Reference Concepts	*Definition*
Absolute location	
Latitude lines	
Longitude Lines	
Prime meridian	

6) Describe the purpose of the Global Positioning System (GPS) and how it connects to absolute location.

7) Describe the purpose of Geographic Information Systems (GIS) and how it connects to research.

8) Using Figures 2.6 and 2.7 (p. 38-39), explain how *remote sensing* images can give us a sense of how economically developed areas of the world are.

LG 2-3: ***What are the consequences of decisions made using geographic data?***

9) Using Figure 2.10 (p. 40), give ONE example of the effects of using Satellite Imagery.

After You Read the Module

Review the Key Terms Part 1

Match each key term on the left with its definition on the right.

____ census	1) The 0° longitude line that runs through Greenwich, England.
____ fieldwork	2) The horizontal line circling earth parallel to the equator, or lines that run north or south of the equator.
____ absolute location	3) The vertical lines on earth surface that mark imaginary circles, connecting the North Pole with the South Pole.
____ latitude lines	4) An official count or survey of a population, typically recording various details about individuals, such as age, sex, race, and location.
____ longitude lines	5) Learning and doing research involve firsthand experience, which takes place outside the classroom setting.
____ Prime Meridian	6) A precise position on Earth's surface.

Review the Key Terms Part 2

Match each key term on the left with its definition on the right.

____ Global Positioning System	1) The scanning of Earth by satellite or high-flying aircraft to obtain information about it.
____ geographic information System	2) Images of the Earth's surface gathered from sensors mounted on orbiting satellites.
____ remote sensing	3) A system of 24 satellites that orbit Earth twice daily and transmit radio signals Earthward.
____ aerial photography	4) Remote-sensing photography that produces fine-grained, high-resolution, highly detailed images.
____ satellite imagery	5) A software application for capturing, storing, checking, and displaying data related to positions on Earth's surface.

Module 3
Spatial Concepts and Types of Diffusion

Before You Read the Module

Focus on Learning Goals

Use the Module learning goals as a way to divide or "chunk" the modules into parts. After you read a section of the module and answer the questions in this book, you should stop and ask yourself: "do I sufficiently understand the goal?"

LG 3-1: Explain geographical concepts used to analyze connections between places and people.

LG 3-2: Describe the different types of diffusion.

Key Terms

Take a look at the key terms you will encounter in this module.

relative location
place
time-space compression
geographic processes
independent invention
hierarchical diffusion
contagious diffusion
relocation diffusion
space
cultural landscape
interdependence
diffusion
expansion diffusion
reverse hierarchical diffusion
stimulus diffusion
friction of distance

While You Read the Module

Answer the questions as you read Module 3: Spatial Concepts and Types of Diffusion.

LG 3-1: ***What geographical concepts are used to analyze connections between places and people?***

1) Complete the following table.

Type of Location	*Definition*	*Example*
Absolute location		
Relative location		

2) Describe how the concepts of space and place are different to human geographers.

3) Define *cultural landscape*.

4) Using Figure 3.2 (p. 46), describe the pagoda's design and identify what religion would have a special connection to this place.

5) Define *time-space compression* and give a specific example of how "the world is shrinking."

6) Globalization has caused us to become more interdependent and connected. According to the book, what is one impact of this on young people in Rwanda?

7) Define *geographic processes*.

LG 3-2: *Describe the different types of diffusion.*

8) Complete the following table to explain the two different ways different places are similar.

Term	*Definition*
Independent invention	
Diffusion	

9) Define *expansion diffusion* and identify the 3 types of expansion diffusion

10) Define *relocation diffusion*.

11) Using Figure 3.5 (p. 50), draw the types of diffusion on your own below.

Type of Diffusion	*Drawing of each*
Relocation Diffusion	
Hierarchical Expansion Diffusion	
Contagious Expansion Diffusion	

After You Read the Module

Review the Key Terms Part 1

Match each key term on the left with its definition on the right.

____ relative location	1) A pattern by which a phenomenon, such as the movement of people, or their ideas, technologies, or preferences, spreads from a particular location through space and time.
____ space	2) How we modify space based on who we are as a group of people.
____ place	3) The built forms that cultural groups create in inhabiting Earth.
____ cultural landscape	4) The areas we occupy as humans; it has no value until the people who occupy it, make it their own.
____ time space compression	5) The decreasing distance between places, as measured by travel time or cost.
____ interdependence	6) The physical and human forces that work together to form and transform the world.
____ geographic processes	7) Occurs when the same or very similar innovation is developed at the same time, in a different place, by different people working independently.
____ diffusion	8) The ties established between regions and countries that, over time, collectively create a global economic system that is not necessarily based on equality.
____ independent invention	9) The position of one place (or person) about the position of another place (or person).

Review the Key Terms Part 2

Match each key term on the left with its definition on the right.

____ expansion diffusion

____ hierarchical diffusion

____ reverse hierarchical diffusion

____ contagious diffusion

____ stimulus diffusion

____ relocation diffusion

____ friction of distance

1) The wavelike spread of ideas in the manner of a contagious disease or forest fire, moving throughout space without regard for hierarchy.
2) Occurs when ideas leapfrog from a lower level of a hierarchy to a higher level.
3) The inhibiting effect of distance on the intensity and volume of most forms of human interaction.
4) Occurs when ideas leapfrog from one important person, community, or city to another, bypassing other persons, communities, or rural areas.
5) Occurs when individuals or groups with a particular idea or practice migrate from one location to another, thereby bringing the idea or practice to a new homeland.
6) Occurs when a specific trait is rejected, but the underlying idea is accepted.
7) Occurs when ideas or practices spread throughout a population, from area to area, in a snowballing process, so that the total number of numbers or users and the areas of occurrence increase.

Module 4
Human-Environmental Interaction

Before You Read the Module

Focus on Learning Goals

Use the Module learning goals as a way to divide or "chunk" the modules into parts. After you read a section of the module and answer the questions in this book, you should stop and ask yourself: "do I sufficiently understand the goal?"

LG 4-1: Explain how geographers view the interaction between people and the natural environment.

LG 4-2: Explain the evolution of geographic thought.

Key Terms

Take a look at the key terms you will encounter in this module.

ecology	cultural ecology
ecosystem	environmental perception
natural hazard	tsunami
natural resources	non-renewable resources
renewable resources	greenhouse gases
greenhouse effect	environmental determinism
possibilism	

While You Read the Module

Answer the questions as you read Module 4: Human-Environmental Interaction.

LG 4-1: ***How do geographers view the interaction between people and the natural environment?***

1) Ecology is the biological science concerned with studying complex relationships among living organisms and their physical environments. Describe what *cultural ecology* is.

2) Using Figures 4.1-4.5 (p. 54-55), complete the table below by describing the human-environmental interaction.

Figures 4.1-4.5	*Describe the Human-Environmental Interaction*
4.1	
4.2	

4.3	
4.4	
4.5	

3) Humans' environmental perceptions of natural hazards help prepare against natural disasters. Using your own local natural environmental landscape at your school, identify ONE natural hazard where you live and describe how people have prepared in case of an emergency.

4) On page 56, the text discusses "Japan's so-called triple disaster." Describe how people prepared for each of these. Was the planning effective or ineffective?

Triple disaster	*Planning and Impact*
Earthquake	
Tsunami	
Flooded Nuclear Power Plant	

5) Describe the differences between non-renewable resources and renewable resources, and give ONE example of each type of resource.

6) Define *greenhouse gases* and describe ONE impact of the increased usage of greenhouse gases on Earth.

LG 4-2: *How has geographic thought evolved?*

7) Two theories have been used historically to explain the human-environmental interaction. Define each theory and explain how geographic thinking has evolved, leading to one theory's acceptance today.

After You Read the Module

Review the Key Terms Part 1

Match each key term on the left with its definition on the right.

____ ecology	1) A territorial bounded system consisting of the interaction between humans and the environment.
____ cultural ecology	2) The mental images that comprise humans' perception of nature.
____ ecosystem	3) A biological science concerned with studying the complex relationship among living organisms and their physical environments.
____ environmental perception	4) The study of the interactions between societies and their local environments.
____ natural hazard	5) A huge ocean wave produced by the displacement of a large volume of water, often caused by an earthquake.
____ tsunami	6) A physical danger present in the environment.

Review the Key Terms Part 2

Match each key term on the left with its definition on the right.

____ natural resources	1) The belief that any physical environment offers several possible ways for society to develop, and that humans can find ways to overcome environmental challenges.
____ non-renewable resources	2) The global warming trend is caused by rising levels of carbon dioxide.
____ renewable resources	3) The belief that their physical environment is the dominant force shaping cultures, and that humanity is a passive product of its physical surroundings.
____ greenhouse gases	4) Natural resources that are available on Earth in finite quantities.
____ greenhouse effect	5) compounds in the atmosphere from fossil-fuel combustion.
____ environmental determinism	6) Natural resources that Earth will naturally replenish over time.
____ possibilism	7) Materials, or substances that occur in nature and can be used for economic gain.

Module 5
Scales of Analysis

Before You Read the Module

Focus on Learning Goals

Use the Module learning goals as a way to divide or "chunk" the modules into parts. After you read a section of the module and answer the questions in this book, you should stop and ask yourself: "do I sufficiently understand the goal?"

LG 5-1: Define geographic scale of analysis.

LG 5-2: Describe how geographers apply scale in geographic analysis.

Key Terms

Take a look at the key terms you will encounter in this module.

global scale
regional scale analysis
national scale analysis
local scale analysis
glocal perspective

While You Read the Module

Answer the questions as you read Module 5: Scales of Analysis.

LG 5-1: ***What is a geographic scale of analysis?***

1) Identify the four most common types of geographic scales of analysis.

2) Using Figure 5.2 (p. 68), identify areas of large concentrations of heavy metal pollution across the globe using a global scale.

3) Using Figure 5.3 (p. 69), describe the difference between the average wealth of countries in the EU versus non-EU countries.

4) Using Figure 5.4 (p. 70), identify the two German states that have the highest per capita GDP.

LG 5-2: ***How do geographers apply scale in geographic analysis?***

5) Using Figure 5.5 (p. 71), identify states that had the highest population growth rates from 2010 and 2020.

6) Using Figure 5.7 (p. 73), identify the percentage of Buddhists in China.

7) Using Figure 5.7 (page 73), describe the distribution of Buddhists in China: clustered or dispersed.

8) After analyzing Figure 5.7 (p. 73), explain how looking at different scales of analysis helps us have a better understanding of spatial patterns.

9) Describe how Burger King in South Korea is a good example of glocalization (the glocal perspective).

After You Read the Module

Review the Key Terms

Match each key term on the left with its definition on the right.

____ global scale	1) Geographic scale that identifies and analyzes geographic phenomena within a state or province, a city or town, or a neighborhood.
____ regional scale analysis	2) Geographic scale that identifies and analyzes geographic phenomena within a specific country.
____ national scale analysis	3) Geographic scale that looks at geographic phenomena across the entire world.
____ local scale analysis	4) Geographic perspective that acknowledges the two-way relationship between local communities and global patterns.
____ global perspective	5) Geographic scale that identifies and analyzes geographic phenomena within a particular region.

Module 6
Regional Analysis

Before You Read the Module

Focus on Learning Goals

Use the Module learning goals as a way to divide or "chunk" the modules into parts. After you read a section of the module and answer the questions in this book, you should stop and ask yourself: "do I sufficiently understand the goal?"

LG 6-1: Describe spatial patterns using characteristics of regions.

LG 6-2: Explain how region regional boundaries are evolving.

Key Terms

Take a look at the key terms you will encounter in this module.

region
border zone
nodes
perceptual/vernacular region
sense of place
regional identity
regional analysis
formal region
functional region
metropolitan area
mental map
activity space
contested boundaries

While You Read the Module

Answer the questions as you read Module 6: Regional Analysis.

LG 6-1: ***How would you describe spatial patterns, using characteristics of regions?***

1) Identify the three types of regions you will learn about in this module.

2) Define region

3) Understanding the three types of regions will be important throughout the year. Complete the following table below and refer to it as the year progresses.

Type of Region	*Definition*	*Example*
Formal Region		
Function Region		
Perceptual/ Vernacular Region		

4) Explain why the shaded areas on the map in Figure 6.1 (p. 79) are considered a formal region.

5) Functional regions have nodes. Define a node and describe a metropolitan area using the term "node" in your response.

6) Describe the photo in Figure 6.3 (p. 80) and what type of region it must be based on the characteristics.

7) Explain why regions like "the South", "the Midwest", "the Middle East", or "South Asia" are perceptual/vernacular regions.

8) Fill in the table for the terms associated with the importance of mental maps.

Terms	*Definition*
Mental Map	
Sense of Place	
Activity Space	

LG 6-2: *Changing Regional Boundaries*

9) Regional identities can happen across multiple scales, histories, and ancestries. Fill out the table on how you would identify yourself based on the following situations.

Where are you?	*Identity you would give to others*
Far away from your home country on vacation	
In a different region within your country	
In a different town close to home	

10) Define *contested boundaries*.

11) Using Figure 6.8 (p. 84), describe how the Kurdish population boundaries create contested boundaries for the countries in the region.

12) Define *regional analysis*.

After You Read the Module

Review the Key Terms Part 1

Match each key term on the left with its definition on the right.

____ region	1) An area, composed of a heavily populated, urban core and its less populated surrounding areas.
____ formal region	2) A geographical area that is perceived to exist by its inhabitants, based on widespread acceptance and the use of a unique regional.
____ border zone	3) The central point where the functions of a functional region are coordinated and directed.
____ functional region	4) A geographical area that has been organized to function, politically, socially, culturally, or economically as one unit.
____ nodes	5) A region where cultural markers overlap and blend in a recognizable border culture.
____ metropolitan area	6) A geographical unit based on one or more common characteristics or functions.
____perceptual/vernacular region	7) A geographical area inhabited by people who have one or more traits in common.

Review the Key Terms Part 2

Match each key term on the left with its definition on the right.

____ mental map	1) How a person feels about a particular place and why it's important to them.
____ sense of place	2) The process of examining patterns and processes within and between regions at multiple geographic scales.
____ activity space	3) A personal representation of a portion of the Earth's surface.
____ regional identity	4) Where a person goes and what she does here on a day-to-day basis.
____ contested boundaries	5) The awareness of belonging to a group of people within a region.
____ regional analysis	6) Boundaries that are disputed for religious, political, or cultural reasons.

Unit 1
Review Exercises

Check Your Understanding

Review the "Module Review" summaries at the end of each module in Unit 1of your textbook. Make sure you understand all of the concepts before moving on to the next unit.

Unit 1 Multiple-Choice Question AP® Exam Practice

1. Which of the following best explains the impact of time-distance decay on human interactions?
(A) It increases the importance of absolute distance in determining migration patterns.
(B) It suggests that places closer together are more likely to have stronger interactions due to reduced friction of distance.
(C) It argues that cultural traits diffuse at the same rate regardless of distance.
(D) It is only applicable to transportation networks, not cultural diffusion.
(E) It contradicts the idea of spatial perspective by minimizing the importance of geography.

2. Why is remote sensing particularly valuable for understanding economic development?
(A) It allows researchers to conduct interviews with local business owners.
(B) It provides high-resolution images of urban infrastructure and nighttime light emissions, which correlate with economic activity.
(C) It is the primary method used to measure absolute location.
(D) It replaces the need for fieldwork by providing all necessary demographic data.
(E) It ensures complete accuracy in determining a country's GDP.

3. Which of the following best describes the relationship between census data and political representation in the United States?
(A) Census data determine the number of electoral votes each state receives in presidential elections.
(B) Census data directly influence state laws regarding voter registration.
(C) Census data are used to determine the number of seats each state has in the U.S. House of Representatives.
(D) Census data dictate how governors are elected in each state.
(E) Census data establish the political party representation within the Senate.

4. Which of the following best characterizes the concept of cultural ecology?
(A) It focuses exclusively on biological processes in natural ecosystems without considering human behavior.
(B) It examines how human cultural practices both shape and are shaped by the local natural environment.
(C) It is primarily concerned with measuring the economic value of natural resources.
(D) It argues that human societies are completely determined by their environment, leaving no

room for cultural innovation.
(E) It rejects the idea that environment influences culture, instead emphasizing cultural evolution in isolation.

5. Which of the following best explains the significance of changing the scale of analysis when examining economic inequality?
   (A) It allows geographers to compare different countries' overall wealth.
   (B) It highlights local disparities that may not be visible at the national or global scale.
   (C) It focuses only on urban areas, ignoring rural economic challenges.
   (D) It eliminates the need for qualitative data in economic research.
   (E) It ensures that all data remain consistent across all geographic levels.

6. Which of the following is an example of a functional region?
   (A) The Corn Belt, defined by its agricultural production.
   (B) The Amazon Rainforest, defined by its ecosystem.
   (C) The metropolitan area of Chicago, connected by transit and economic activity.
   (D) The Middle East, defined by cultural and historical commonalities.
   (E) The Sahara Desert, characterized by its climate.

Unit 1 Free-Response Question AP® Exam Practice

Geographers use tools such as maps, spatial data, and geographic models to understand the ways location, place, scale, and pattern shape human activity and decision-making. Use Figure 1.1. (p. 14) to answer the prompts below. Respond to parts A, B, C, D, E, F, and G.

A. Identify the type of map.
B. Identify the scale of analysis in the map.
C. Describe how the data is aggregated according to the map legend.
D. Describe how contagious diffusion works.
E. Define non-renewable energy and give an example of a non-renewable energy source.
F. Explain how metropolitan regions work as functional regions.
G. Explain a limitation of the map in understanding the number of people who live in each state.

Notes:

Unit 2
Population and Migration Patterns and Processes

Unit Summary

The study of population provides a crucial foundation for understanding many other topics in human geography, such as cultural diffusion and change, shifts in political landscapes, rural land use and food production, urbanization, natural resource use, as well as economic development. This unit will discuss the fundamental topics in population geography, including population distribution, composition, growth, and migration. It will also cover the demographic transition model, the Malthusian theory, women's role in demographic change, population policy issues, and population aging, which is a challenge confronting a growing number of countries.

Modules in Unit 2

Module 7: Population Distribution and Its Consequences
Module 8: Population Composition
Module 9: Population Dynamics
Module 10: The Demographic Transition Model
Module 11: Malthusian Theory
Module 12: Population Policies
Module 13: Women and Demographic Change
Module 14: Aging Populations
Module 15: Causes of Migration
Module 16: Forced and Voluntary Migration
Module 17: Effects of Migration

What do you see?

Look closely at the unit-opening photograph. Write down 5 things you observe about it. You will return to this image in the AP® Seeing Geography feature at the end of the unit.

1)

2)

3)

4)

5)

Module 7
Population Distribution and Its Consequences

Before You Read the Module

Focus on Learning Goals

Use the Module learning goals as a way to divide or "chunk" the modules into parts. After you read a section of the module and answer the questions in this book, you should stop and ask yourself: "do I sufficiently understand the goal?"

LG 7-1: Describe how population is distributed around the world.
LG 7-2: Identify the factors that influence population distribution.
LG 7-3: Explain the three methods for calculating population density.
LG 7-4: Explain the consequences of population distribution and density.

Key Terms

Take a look at the key terms you will encounter in this module.

population distribution
Eurasia
ecumene
population clusters
metacity
megacity
developed (or industrialized) country
developing (or industrializing) countries
Snow Belt
Sunbelt
mean center of population
population density
arithmetic (crude) density
physiological density
arable land
agricultural density
carrying capacity
human well-being

While You Read the Module

Answer the questions as you read Module 7: Population Distribution and Its Consequences.

LG 7-1: ***How is population distributed around the world?***

1) Define *population distribution.*

2) Describe the unevenness in global population distribution.

3) Define *Eurasia.*

4) Define ecumene.

5) Describe the world's four major population clusters

6) Describe Snow Belt and Sunbelt.

7) Describe how the mean center of population in the United States has shifted over time.

LG 7-2: ***What factors influence population distribution?***

8) Explain how physical factors such as elevation and water bodies affect population distribution.

9) Explain how climate affects population distribution.

10) Explain how culture may affect population distribution.

11) Describe the effect of economic development on population distribution.

12) Explain how diseases may affect population distribution

LG 7-3: ***What are the three methods for calculating population density?***
13) Define *population density.*

14) Complete the following table.

Density Indicators	*Definition*
Arithmetic (crude) density	
Physiological Density	
Agricultural Density	

15) Define *arable land.*

16) Explain why developed countries usually have lower agricultural density than developing countries.

LG 7-4: ***What are the consequences of population distribution and density?***
17) Describe the effects of population distribution on economic development.

18) Explain how changes in population distribution can affect political power distribution in the United States.

19) Define *carrying capacity*.

20) Describe the relationship between population distribution and human well-being.

21) Explain how population distribution may affect people's health and vulnerability to natural disasters.

After You Read the Module

Review the Key Terms Part 1

Match each key term on the left with its definition on the right.

____ population distribution	1) The portion of Earth's surface with permanent human settlement.
____ ecumene	2) Heavily populated areas that illustrate the unevenness in global population distribution.
____ Eurasia	3) A city with more than 10 million residents.
____ population clusters	4) A country with an advanced economy and a high standard of living.
____ metacity	5) The pattern in which humans are spread out on Earth's surface.
____ megacity	6) A city with more than 20 million residents.
____ developed (or industrialized) country	7) Countries that are of relatively low income or economically poorer than developed countries.

____ developing (or industrializing) countries | 8) A massive piece of land on Earth that consists of Europe, with less than 10 percent of the human population, and Asia, which accounts for just under 60 percent of humanity.

Review the Key Terms Part 1

Match each key term on the left with its definition on the right.

____ Snow Belt	1) The balancing point given the distribution of population.
____ Sunbelt	2) The average number of people per unit of land area (usually per square mile or kilometer).
____ mean center of population	3) The number of people a particular environment or Earth as a whole can support on a sustainable basis.
____ population density	4) The average number of people per unit area (a square mile or kilometer) of arable land.
____ arithmetic (crude) density	5) States in coastal areas and the South and Southwest of the United States.
____ physiological density	6) The state of being comfortable, healthy, or happy.
____ arable land	7) The number of farmers per unit of arable land.
____ agricultural density	8) States located in the northern and midwestern parts of the United States.
____ carrying capacity	9) The average number of people per unit of land area.
____ human well-being	10) Land suitable for cultivation.

Module 8
Population Composition

Before You Read the Module

Focus on Learning Goals

Use the Module learning goals as a way to divide or "chunk" the modules into parts. After you read a section of the module and answer the questions in this book, you should stop and ask yourself: "do I sufficiently understand the goal?"

LG 8-1: Describe the age structure of a population.
LG 8-2: Identify major social generations in the United States.
LG 8-3: Define the sex ratio of a population.
LG 8-4: Explain the uses of population pyramids.

Key Terms

Take a look at the key terms you will encounter in this module.

population composition
age structure
dependency ratio
youth dependency ratio
elderly dependency ratio
generations
baby boomers
Generation X
Millennials
Generation Z
sex ratio
androcentrism
infanticide
population pyramid

While You Read the Module

Answer the questions as you read Module 8: Population Composition.

LG 8-1: ***How do we describe the age structure of a population?***

1) Define *population composition.*

2) Define *age structure.*

3) Describe dependency ratios.

4) Identify a country with high youth dependency ratio.

5) Identify a country with high elderly dependency ratio

LG 8-2: ***What are the major social generations in the United States?***

6) Define *generations*.

7) Complete the following table.

Social Generations	*Definition*
Baby Boomers	
Generation X	
Millennials	
Generation Z	

8) Using Figure 8.3 (p. 122), identify another baby boom in the United States since the 1960s.

9) Describe the characteristics of the Millennial generation.

LG 8-3: ***What is the sex ratio of a population?***

10) Define *sex ratio*.

11) Define *androcentrism*.

12) Define *infanticide*.

13) Several factors may cause unbalanced sex ratio in a population. Complete the following table.

Factor	*Describe how the factor affects sex ratio*
Cultural preferences for males	
Gender-selective migration	
War	

14) Describe potential consequences of unbalanced sex ratios.

LG 8-4: ***What can a population pyramid be used for?***

15) Describe population pyramids.

16) The shape of population pyramids reflects past population growth. Complete the following table.

Population Growth	*Describe the shape of the population pyramid*
Rapid growth	
Slow growth	
Stable population	
Negative growth (Declining population)	

17) Define *thematic map*.

18) Explain how a country's population pyramid may be used to predict future needs for goods and services.

19) Explain how a country's population pyramid may be used to predict the future size of the labor force.

20) Describe the typical shape of population pyramid for a college town.

21) Describe the likely shape of population pyramid for a retirement community.

After You Read the Module

Review the Key Terms

Match each key term on the left with its definition on the right.

____ population composition	1) The number of dependents in a population that every 100 working-age people (ages 15 to 64 years) must support.
____ age structure	2) People who were born between 1981 and 1996.
____ dependency ratio	3) Refers to the breakdown of a population into different age groups or cohorts.

_____ youth dependency ratio

_____ elderly dependency ratio

_____ generations

_____ baby boomers

_____ Generation X

_____ Millennials

_____ Generation Z

_____ sex ratio

_____ androcentrism

_____ infanticide

_____ population pyramid

4) Groups of people who were born around the same time and share some common traits due to the cultural and societal influences they shared as they grew up.

5) The makeup of a population by age and sex as well as by ethnic, racial, income, and educational background.

6) A very useful graphic device for depicting the age and sex structure of a population.

7) The ratio of the number of men to number of women in a population.

8) The number of young dependents in a population (usually people younger than 15 years of age) that every 100 working-age people must support.

9) People born between 1965 and 1980 and who are now in their prime working years.

10) A phenomenon in which a culture demonstrates a marked preference for males.

11) The practice of killing infants.

12) People born between 1997 and 2012.

13) People born from 1946 to 1964 during the post–World War II uptick in birth rate.

14) The number of elderly dependents in a population (usually people older than 64 years of age) that every 100 working-age people must support.

Module 9
Population Dynamics

Before You Read the Module

Focus on Learning Goals

Use the Module learning goals as a way to divide or "chunk" the modules into parts. After you read a section of the module and answer the questions in this book, you should stop and ask yourself: "do I sufficiently understand the goal?"

LG 9-1: Explain how the world's population has grown over time.
LG 9-2: Explain how we measure birth rates.
LG 9-3: Explain how we measure death rates.
LG 9-4: Explain how we measure natural increase.

Key Terms

Take a look at the key terms you will encounter in this module.

demographic equation
crude birth rate (CBR)
low birth rate
transitional birth rate
high birth rate
total fertility rate (TFR)
replacement level fertility
gender roles
crude death rate (CDR)
infant mortality rate (IMR)
child mortality
rate of natural increase (RNI)
zero population growth
doubling time
rule of 70

While You Read the Module

Answer the questions as you read Module 9: Population Dynamics.

LG 9-1: ***How has the world's population grown over time?***

1) Describe the demographic equation.

2) Based on the demographic equation, describe one factor that may lead to an increase in total population in your county or city.

3) Based on the demographic equation, describe one factor that may lead to a decrease in total population of a developed country.

4) Describe world population growth since the Industrial Revolution in the mid-eighteenth century.

5) Identify the world regions or countries that will contribute much of the projected increase in global population up to 2050.

LG 9-2: ***How do we measure birth rates?***

6) Define *crude birth rate (CBR)*.

7) Complete the following table.

Level of Birth Rate	*Definition*
Low Birth Rate	
Transitional Birth Rate	
High Birth Rate	

8) Define *total fertility rate (TFR)*.

9) Describe replacement level fertility.

10) Explain how current total fertility rates vary by world region.

11) A number of external factors influence total fertility rate (TFR). Complete the following table.

Factor	*Effects*
Economic Development	
Religion and Culture	
Education and Women's Employment	
Gender Roles	
Population Policies	

LG 9-3: *How do we measure death rates?*

12) Define *crude death rate (CDR).*

13) Explain how a population's age-sex structure and economic development level affect crude death rate (CDR).

14) Describe the trends and patterns in death rates worldwide.

15) Define *infant mortality rate (IMR).*

16) Using Figure 9.6 (p. 138), identify a world region with the highest infant mortality rate (IMR) and a world region with a low infant mortality rate.

17) Explain why experts believe that infant mortality rate may be the best single indicator of living standards.

18) Define *child mortality*.

LG 9-4: *How do we measure natural increase?*

19) Define the *rate of natural increase (RNI)*.

20) Using Figure 9.7 (p. 139), identify the world region with the highest RNI and a world region with low or negative RNI.

21) Describe one difference between TFR and RNI as indicators of population growth.

22) Describe how a country may have zero population growth (ZPG).

23) Define doubling time of a population.

24) Describe the rule of 70 for calculating a population's doubling time.

25) Explain how developed and developing countries may differ in doubling time.

After You Read the Module

Review the Key Terms

Match each key term on the left with its definition on the right.

____ demographic equation	1) The difference between the number of births and deaths in a given year, when expressed as a percentage of total population.
____ crude birth rate (CBR)	2) The number of years it takes for a population to double in size.
____ low birth rate	3) Behaviors and attitudes associated with a particular gender based on cultural norms.
____ transitional birth rate	4) The average number of children needed to replace both parents and stabilize population over time.
____ high birth rate	5) The average number of births per 1000 people; the traditional way of measuring birth rates.
____ total fertility rate (TFR)	6) Deaths of children under five years of age.
____ replacement level fertility	7) When a country has the same number of births and deaths in a given year, its RNI is zero.
____ gender roles	8) A tool for calculating the doubling time of a population by dividing 70 by a country's rate of natural increase (RNI).
____ crude death rate (CDR)	9) The method for calculating total population of a country or place based on natural increase and migration over a period of time (usually a year).
____ infant mortality rate (IMR)	10) A crude birth rate below 20 births per 1000 people.

____ child mortality

11) A crude birth rate of more than 30 per 1000 people.

____ rate of natural increase (RNI)

12) A measure of how many infants die within the first year of their life per 1000 live births.

____ zero population growth (ZPG)

13) A crude birth rate between 20 and 30 per 1000 people.

____ doubling time

14) The average number of children born per woman during her reproductive lifetime, considered to be from 15 to 49 years of age.

____ rule of 70

15) The number of deaths per year per 1000 people.

Module 10
The Demographic Transition Model

Before You Read the Module

Focus on Learning Goals

Use the Module learning goals as a way to divide or "chunk" the modules into parts. After you read a section of the module and answer the questions in this book, you should stop and ask yourself: "do I sufficiently understand the goal?"

LG 10-1: Explain how birth and death rates change over time.
LG 10-2: Discuss whether the demographic transition model still applies today.
LG 10-3: Explain the epidemiological transition model.

Key Terms

Take a look at the key terms you will encounter in this module.

demographic transition model (DTM)
demographic dividend
epidemiology
epidemiological transition theory
degenerative disease

While You Read the Module

Answer the questions as you read Module 10: The Demographic Transition Model.

LG 10-1: ***How do birth and death rates change over time?***

1) Describe the demographic transition model (DTM).

2) The DTM includes five stages or phases. Complete the following table.

Stages	*Demographic Characteristics*
1: High Stationary Stage	
2. Early Expanding Stage	
3. Late Expanding Stage	
4. Low Stationary Stage	

5. Natural Decrease Stage	

3) Define *demographic dividend.*

LG 10-2: ***Does the demographic transition model still apply today?***

4) Describe one complication in apply the DTM in developing countries today.

5) Explain how changes in some countries' birth and death rates may not follow the model's predictions.

LG 10-3: ***What is the epidemiological transition model?***

6) Define *epidemiology*.

7) Describe the epidemiological transition theory.

8) The epidemiological transition theory identifies four phases. Complete the following table.

Phase	*The Characteristics*
The Age of Pestilence and Famine	
The Age of Receding Pandemics	

The Age of Degenerative and Human-Made Diseases	
The Age of Delayed Degenerative Diseases	

9) Define *degenerative diseases*.

10) Identify the top 5 causes of death in low-income countries.

11) Identify the top 5 causes of death in high-income countries.

After You Read the Module

Review the Key Terms

Match each key term on the left with its definition on the right.

____ demographic transition model (DTM)	1) Seeks to explain how changes in health services and living standards affect patterns of disease.
____ demographic dividend	2) A branch of medicine that studies the distribution, determinants, and control of diseases and other health conditions, such as tobacco use and sedentary lifestyle.
____ epidemiology	3) A disease that causes deterioration over time, such as cancer, heart disease, and stroke.
____ epidemiological transition theory	4) Conceptualizes how crude birth rate (CBR) and crude death rate (CDR) as well as the resulting rate of natural increase (RNI) change over time as countries go through industrialization and urbanization.
____ degenerative disease	5) An economic benefit countries get when the share of their working-age population rises relative to children and elderly people.

Module 11
Malthusian Theory

Before You Read the Module

Focus on Learning Goals

Use the Module learning goals as a way to divide or "chunk" the modules into parts. After you read a section of the module and answer the questions in this book, you should stop and ask yourself: "do I sufficiently understand the goal?"

LG 11-1: Describe the Malthusian theory of population.
LG 11-2: Explain the criticisms of the Malthusian theory.
LG 11-3: Explain the carrying capacity of Earth.

Key Terms

Take a look at the key terms you will encounter in this module.

Malthusian
Malthusian Trap
Overpopulation
Neo-Malthusians
Cornucopians or anti-Malthusians
Boserup effect

While You Read the Module

Answer the questions as you read Module 11: Malthusian Theory.

LG 11-1: ***What is the Malthusian theory of population?***

1) Define *Malthusian*.

2) Compare the difference between geometric growth and arithmetic growth.

3) Describe Malthusian Trap.

4) Define *overpopulation*.

5) Explain the positive check in the Malthusian theory and why it is not a good solution for population reduction.

6) Describe preventive or negative checks in the Malthusian theory.

7) Define *neo-Malthusians*.

LG 9-2: *What are the criticisms of the Malthusian theory?*

8) Define *Cornucopians* or *anti-Malthusians*.

9) Describe the Cornucopian argument against the Malthusian theory.

10) Define the *Boserup effect*.

11) Explain how the Malthusian predictions about food production and population growth fared against real world experiences?

12) Using Figure 11.7 (p. 159), identify a country with more than 30 percent of its population still using an unimproved drinking water source in 2022.

13) Using Figure 11.8 (p. 159), identify the world region with only a small proportion of the population using improved sanitation facilities.

LG 9-3: ***What is the carrying capacity of Earth?***

14) Explain why it is difficult to determine carrying capacity for a particular region or the whole world.

15) Describe one way the carrying capacity of a place can be expanded.

After You Read the Module

Review the Key Terms

Match each key term on the left with its definition on the right.

____ Malthusian	1) People who reject the Malthusian view of population and resources.
____ Malthusian Trap	2) Increase in food production resulting from the use of new farming methods.
____ overpopulation	3) Population growth outpaces food production, leading to natural resource depletion, starvation, poverty, violence, and population decline.
____ neo-Malthusians	4) A term derived from the name Thomas Robert Malthus, an English economist and cleric, to mean either "of or relating to Malthus's theory" or "a follower of Malthus".
____ cornucopians or anti-Malthusians	5) Occurs when the human population exceeds the food supply.
____ Boserup effect	6) People who today subscribe to the Malthusian view of population.

Module 12
Population Policies

Before You Read the Module

Focus on Learning Goals

Use the Module learning goals as a way to divide or "chunk" the modules into parts. After you read a section of the module and answer the questions in this book, you should stop and ask yourself: "do I sufficiently understand the goal?"

LG 12-1: Describe antinatalist policies and their purpose.
LG 12-2: Describe pronatalist policies and their purpose.

Key Terms

Take a look at the key terms you will encounter in this module.

antinatalist policies
pronatalist policies

While You Read the Module

Answer the questions as you read Module 12: Population Policies.

LG 12-1: ***What are antinatalist policies and what is their purpose?***

1) Define *antinatalist policies*.

2) Describe how China's birth control policies have changed since the 1970s.

3) Identify the social and demographic consequences of China's one couple one child policy.

4) Identify another example of antinatalist policy implemented by a country other than China.

5) Explain why it may be easier to implement antinatalist policies in non-democratic countries.

LG 12-2: ***What are pronatalist policies and what is their purpose?***

6) Define pronatalist policies.

7) Identify common incentives and social programs used by governments to increase fertility rates.

8) Explain why pronatalist policies often have limited effects on fertility.

After You Read the Module

Review the Key Terms

Match each key term on the left with its definition on the right.

_____ antinatalist policies	1) Policies designed to boost fertility rates and ultimately population growth.
_____ pronatalist policies	2) Policies designed to curtail population growth by reducing fertility rates.

Module 13
Women and Demographic Change

Before You Read the Module

Focus on Learning Goals

Use the Module learning goals as a way to divide or "chunk" the modules into parts. After you read a section of the module and answer the questions in this book, you should stop and ask yourself: "do I sufficiently understand the goal?"

LG 13-1: Describe the concept of women's status.
LG 13-2: Explain how women's status affects fertility rates.
LG 13-3: Explain how women's status affects mortality rates.
LG 13-4: Explain how women's status affects migration.

Key Terms

Take a look at the key terms you will encounter in this module.

women's status
women's empowerment

While You Read the Module

Answer the questions as you read Module 13: Women and Demographic Change.

LG 13-1: ***What does women's status mean?***

1) Define *women's status.*

2) Define *women's empowerment.*

LG 13-2: ***How does women's status affect fertility rates?***

3) Explain how women's education affects fertility.

4) Explain how women's access to family planning services affect fertility.

5) Explain how women's participation in economic activities affects fertility.

LG 13-3: ***How does women's status affect mortality rates?***

6) Describe the effects of women's status on mortality rates.

LG 9-4: ***How does women's status affect migration?***

7) Explain how women's status affects migration.

After You Read the Module

Review the Key Terms

Match each key term on the left with its definition on the right.

____ women's status	1) The increased autonomy of women to make choices and shape their lives.
____ women's empowerment	2) The degree of equality between men and women with respect to access to and control over both physical and social resources in the family, community, or society at large.

Module 14
Aging Populations

Before You Read the Module

Focus on Learning Goals

Use the Module learning goals as a way to divide or "chunk" the modules into parts. After you read a section of the module and answer the questions in this book, you should stop and ask yourself: "do I sufficiently understand the goal?"

LG 14-1: Describe the measures and trends of population aging in the world.
LG 14-2: Identify the causes of population aging.
LG 14-3: Identify the consequences of population aging.
LG 14-4: Explain how countries can adapt to aging populations.

Key Terms

Take a look at the key terms you will encounter in this module.

aging population
life expectancy
median age

While You Read the Module

Answer the questions as you read Module 14: Aging Populations.

LG 14-1: ***What are the measures and trends of population aging in the world?***

1) Describe aging population.

2) Define *median age*.

3) Using Figure 14.1 (p. 181), identify the world region with the youngest population and the world region with the oldest population.

4) Using Figure 14.2 (p. 182), identify the world region with the smallest proportion of population 65 years of age and older in 2022.

5) Explain one limitation of the map in Figure 14.2 (p. 182) in showing the geographic patterns of aging.

LG 14-2: ***What are the causes of population aging?***

6) Explain how low fertility rates affect population aging.

7) Define *life expectancy*.

8) Describe the effect of life expectancy on population aging.

9) Describe the geographic pattern of life expectancy at birth as reflected in Figure 14.4 (p. 184).

10) Describe how age-selective migration may contribute to population aging.

LG 14-3: ***What are the consequences of population aging?***

11) Describe the effect of population aging on a country's labor force.

12) Explain how population aging affects elderly dependency ratio.

13) Describe the effects of population aging on a country's economy.

LG 14-4: ***How can countries adapt to aging populations?***

14) Identify what a country can do to boost fertility.

15) Describe one way a country with an aging population can cope with labor shortage.

16) Explain how a country can adapt to an aging society.

After You Read the Module

Review the Key Terms

Match each key term on the left with its definition on the right.

____ aging population	1) The number of years a person can expect to live from birth.
____ median age	2) A population of a country or place that ages as the number or proportion of its elderly people increases.
____ life expectancy	3) The age that divides a population into two halves so that one half is younger than this age and the other half is older.

Module 15
Causes of Migration

Before You Read the Module

Focus on Learning Goals

Use the Module learning goals as a way to divide or "chunk" the modules into parts. After you read a section of the module and answer the questions in this book, you should stop and ask yourself: "do I sufficiently understand the goal?"

LG 15-1: Describe migration.
LG 15-2: Explain the characteristics of migration.
LG 15-3: Explain the push-pull theory.

Key Terms

Take a look at the key terms you will encounter in this module.

spatial mobility
social (upward) mobility
migration
migrant or mover
non-migrants or stayers
origin
destination
out-migration (emigration)
emigrants or out-migrants
in-migration (immigration)
immigrants or in-migrants
migration stream
counterstream
net migration
net migration rate (NMR)
migration age profile
brain drain
brain gain
push-pull theory of migration
push factors
pull factors
intervening obstacles
social networks
intervening opportunity

While You Read the Module

Answer the questions as you read Module 15: Causes of Migration.

LG 15-1: *What is Migration?*

1) Define *spatial mobility.*

2) Define *social (upward) mobility.*

3) Define *migration.*

4) Complete the following table.

Concepts	*Definition*
Migrant or Mover	
Non-migrants or Stayers	
Origin	
Destination	

5) Migrants may move into or out of a place. Complete the following table.

Migration Concepts	*Definition*
Out-migration (Emigration)	
Emigrants or Out-migrants	
In-migration (Immigration)	
Immigrants or In-migrants	

6) Define *migration stream* and *counterstream.*

7) Define *net migration.*

8) Describe how net migration rate (NMR) is calculated.

LG 15-2: ***What are the characteristics of migration?***

9) Describe Ravenstein's laws of migration.

10) Define *migration age profile*.

11) Explain why young people have higher migration rates than other age cohorts.

12) Define *brain drain*.

13) Define *brain gain*.

LG 15-3: ***What is the push-pull theory?***

14) Describe the push-pull theory of migration.

15) Identify three examples of push factors.

16) Identify three examples of pull factors.

17) Describe the role of intervening obstacles in the migration process.

18) Define social networks

19) Describe how intervening opportunity may affect migration.

After You Read the Module

Review the Key Terms

Match each key term on the left with its definition on the right.

____ spatial mobility	1) A phenomenon where a country or a place loses young, more educated, and skilled people through migration.
____ social (upward) mobility	2) People's friends and relatives.
____ migration	3) The act of a migrant arriving at their destination country.
____ migrant or mover	4) The act of a migrant leaving their place (country) of origin.

____ non-migrants or stayers	5) The difference between the number of in-migrants and out-migrants.
____ origin	6) A phenomenon where a country or a place gains young, more educated, and skilled people through migration.
____ destination	7) The flow of all migrants from an origin to a destination.
____ out-migration (emigration)	8) The flow of all migrants in the direction opposite a particular migration stream, from its destination back to the origin.
____ emigrants or out-migrants	9) All forms of geographical movement, including people's everyday commuting and travels.
____ in-migration (immigration)	10) A person who migrates or moves.
____ immigrants or in-migrants	11) The long-term or permanent relocation of individuals, families, or entire communities from one place to another.
____ migration stream	12) The complications that potential migrants will need to overcome to reach their destination.
____ counterstream	13) People who do not move.
____ net migration	14) Mobility that implies a change in socioeconomic status.
____ net migration rate (NMR)	15) A nearby attractive locale where migrants may decide to settle instead of going to the intended destination farther away.

____ migration age profile	16) A person's location before migration.
____ brain drain	17) Factors that cause people to be dissatisfied with their present locales and want to move somewhere else.
____ brain gain	18) The place where the migrant is going.
____ push-pull theory of migration	19) A gauge of the impact of migration on population change, determined by dividing a country's net migration by its total population, then multiplying by 1000.
____ push factors	20) The relatively stable relationship between the odds of migration and age across different countries.
____ pull factors	21) People who leave their country (or place) of origin.
____ intervening obstacles	22) People who arrive at their destination country.
____ social networks	23) Theory asserting that two contrasting sets of factors are at work in migration decisions.
____ intervening opportunity	24) The attributes of other places that make them appealing to potential migrants.

Module 16
Forced and Voluntary Migration

Before You Read the Module

Focus on Learning Goals

Use the Module learning goals as a way to divide or "chunk" the modules into parts. After you read a section of the module and answer the questions in this book, you should stop and ask yourself: "do I sufficiently understand the goal?"

LG 16-1: Describe the types of voluntary migration.
LG 16-2: Explain the mobility transition model.
LG 16-3: Explain forced migration.

Key Terms

Take a look at the key terms you will encounter in this module.

voluntary migration
international migration
guest worker
transnational migration
internal migration or interregional migration
Great Migration
rural-to-urban migration
residential mobility
step (or stepwise) migration
chain migration
return migration
Black Belt
seasonal migration
transhumance
mobility transition model
circulation
forced migration
refugees
internally displaced person (IDP)
asylum seeker
ethnic cleansing
repatriated

While You Read the Module

Answer the questions as you read Module 16: Forced and Voluntary Migration.

LG 16-1: ***What are the types of voluntary migration?***

1) Define *voluntary migration.*

2) Complete the following table.

Type of Migration	*Definition*
International Migration	
Transnational Migration	

Internal (or Interregional) Migration	
Rural-to-Urban Migration	

3) Define *guest worker*.

4) Describe the Great Migration in the United States.

5) Define *residential mobility*.

6) Describe how step (or stepwise) migration works.

7) Describe how chain migration works.

8) Identify two reasons for return migration.

9) Define *Black Belt*.

10) Using Figure 16.4 (p. 207), identify the four major Black Migration streams between 2005 and 2010.

11) Describe two examples of seasonal migration in the United States.

12) Define *transhumance.*

LG 16-2: *What is the mobility transition model?*

13) Define the *mobility transition model.*

14) The mobility transition model includes five stages. Complete the following table.

Stages	*Describe the key characteristics*
Premodern traditional societies	
Early transitional societies	
Late transitional societies	
Advanced societies	
Superadvanced societies	

15) Define *circulation.*

LG 16-3: ***What is forced migration?***

16) Define forced migration.

17) Identify three examples of forced migration.

18) Compare the concept of a refugee with the concept of an internally displaced person (IDP).

19) Define *asylum seeker*.

20) Define *ethnic cleansing*.

21) Identify three examples of ethnic cleansing in the late twentieth and twenty-first centuries.

22) Describe what happens when a refugee is repatriated.

After You Read the Module

Review the Key Terms

Match each key term on the left with its definition on the right.

____ voluntary migration	1) Moves that occur within a metropolitan area.
____ international migration	2) When migrants move back and forth between their home countries and those to which they have migrated.
____ guest worker	3) Migrants going back, or returning, to their previous place of residence or origin.
____ transnational migration	4) When moves are made across national borders.
____ internal migration or interregional migration	5) The twentieth-century movement of 6 million African Americans from the rural southern states to the cities of the midwestern, northeastern, and western states.
____ Great Migration	6) A phenomenon where herders and their livestock move seasonally between their summer and winter pastures.
____ rural-to-urban migration	7) Migration based on the time of year.
____ residential mobility	8) When refugees or displaced persons return to their home country.
____ step (or stepwise) migration	9) Someone who has been forced to leave home because of conflict, disasters, or persecution but stays inside the same country.
____ chain migration	10) Short-term and cyclical movement that occurs repeatedly on a regular basis.
____ return migration	11) A migrant seeking protection from persecution and human rights violations in another country but not yet legally recognized as a refugee.

____ Black Belt	12) When people move within the borders of a country.
____ seasonal migration	13) Geographer Wilbur Zelinsky's conclusion that there are regularities in migration as an essential component of a country's modernization process.
____ transhumance	14) The forced removal of one ethnic group by another ethnic group to create an ethnically uniform territory.
____ mobility transition model	15) People who leave their country because of persecution based on race, ethnicity, religion, nationality, or political opinion.
____ circulation	16) Ethnic homeland of Black Americans in the U.S. South.
____ forced migration	17) When people move from the countryside to cities.
____ refugees	18) Migration caused by forces out of one's control, such as disasters, social conflicts, or developmental projects.
____ internally displaced person (IDP)	19) The process by which some people's migration to a new place leads their family members, friends, and others to move to the same place.
____ asylum seeker	20) A person with temporary permission to work in another country.
____ ethnic cleansing	21) Migration carried out in a series of stages, usually from nearby to bigger and more distant places.
____ repatriated	22) Migration that is done willingly.

Module 17
Effects of Migration

Before You Read the Module

Focus on Learning Goals

Use the Module learning goals as a way to divide or "chunk" the modules into parts. After you read a section of the module and answer the questions in this book, you should stop and ask yourself: "do I sufficiently understand the goal?"

LG 17-1: Describe the geographic effects of migration.

LG 17-2: Explain how international migration creates diaspora cultures.

Key Terms

Take a look at the key terms you will encounter in this module.

remittances

diaspora

While You Read the Module

Answer the questions as you read Module 17: Effects of Migration.

LG 17-1: ***What are the geographic effects of migration?***

1) Explain how migration may affect population characteristics.

2) Explain the economic effects of migration.

3) Define *remittances*.

4) Describe the political effects of migration.

5) Explain how migration can affect the spread of culture, innovations, and diseases.

6) Describe the effect of migration on social cohesiveness.

LG 17-2: ***How does international migration create diaspora cultures?***

7) Define *diaspora*.

After You Read the Module

Review the Key Terms

Match each key term on the left with its definition on the right.

____ remittances	1) Funds sent by international migrants to their families and relatives in their home countries.
____ diaspora	2) Voluntary or involuntary mass dispersions of a population from its home territory.

Unit 2
Review Exercises

Check Your Understanding

Review the "Module Review" summaries at the end of each module in Unit 1 of your textbook. Make sure you understand all of the concepts before moving on to the next unit.

Unit 2 Multiple-Choice Question AP® Exam Practice

1. According to Ravenstein's laws of migration, the main cause of migration is
   (A) Cultural factors
   (B) Economic factors
   (C) Political factors
   (D) Environmental factors
   (E) Social factors

2. Successful early migrants from a rural village often entice their friends and other villagers to move to the same place. Geographers call this phenomenon
   (A) Step migration
   (B) Seasonal migration
   (C) Circulation
   (D) Chain migration
   (E) Transhumance

3. If the current global fertility pattern continues, almost all of the new population growth in the world will come from
   (A) South Asia
   (B) South America
   (C) Subsaharan Africa
   (D) Developing countries
   (E) Developed countries

4. Figure 7.7 (p. 108) shows the mean center of population of the United States has been moving gradually toward southwest. Which of the following best explains the reason for this shift?
   (A) Improving transportation and communication infrastructure
   (B) Large number of tourists visiting southern and southwestern states every year
   (C) Relocation of many companies from California and other states to the South
   (D) Rapid population growth in the Sunbelt states compared to the Snow Belt states
   (E) New methods used by the U.S. Census Bureau to calculate the mean center

5. In most developed countries, the major factor responsible for their future population growth is likely
   (A) Changes in their fertility rates
   (B) Changes in their mortality rates

(C) The number of immigrants they receive
(D) Changes in health care services
(E) Changes in women's status

6. The likely effect of an antinatalist policy on a country's dependency ratios is
   (A) Its youth dependency ratio will increase.
   (B) Its elderly dependency ratio will increase.
   (C) Both youth and elderly dependency ratios will increase.
   (D) Both youth and elderly dependency ratios will decrease.
   (E) An antinatalist policy will have no effect on dependency ratios.

7. A country's population pyramid may be used
   (A) To predict the future needs for goods and services through an analysis of the pyramid shape.
   (B) To estimate the level of emigration out of the country through an analysis of the pyramid shape.
   (C) To estimate rural-to-urban migration within the country through an analysis of the pyramid shape.
   (D) To estimate the infant mortality rate through an analysis of the pyramid shape.
   (E) To estimate the proportion of residents living in urban areas by analyzing the pyramid shape.

8. Based on Figure 9.7 (p. 139), the world region with the highest rate of natural increase currently is
   (A) South America
   (B) Middle America
   (C) South Asia
   (D) North Africa and Southwest Asia
   (E) Subsaharan Africa

9. Based on Figure 10.3 (p. 146), which state of the demographic transition is China likely in?
   (A) Stage 3
   (B) Stage 4
   (C) Stage 5
   (D) Moving from Stage 2 to Stage 3
   (E) Moving from Stage 3 to Stage 4

10. Thomas Malthus claims that population would grow at a geometric rate, while the means of subsistence grow at
    (A) A steadily low rate
    (B) A declining rate
    (C) A variable rate depending on soil and climate
    (D) An arithmetic rate
    (E) An unstable rate

11. South Korea's total fertility rate in 2024 was only 0.7, which means
    (A) South Korea's birth rate is 7 per 1000 people.
    (B) South Korea's population is growing at a rate of 0.7 percent per year.
    (C) 70 percent of South Korea's women are in their reproductive lifetime, from 15 to 49 years of age.
    (D) 7 out of every 10 women in South Korea have children.

(E) The average number of children born per woman during her reproductive lifetime is 0.7.

12. Based on the data shown in Figure 14.1 (p. 181), countries in which part of the world have the greatest potential to benefit from the demographic dividend because of their youthful population?
(A) East Asia
(B) South America
(C) Africa
(D) South Asia
(E) South East Asia

Unit 2 Free-Response Question AP® Exam Practice

Figure 14.2 (p. 182) shows percentages of people ages 65 and older by country. Respond to parts A, B, C, D, E, F, and G.

A. Identify the world region with the smallest proportion of population ages 65 and older.
B. Describe one factor that causes the aging of a population.
C. Describe one likely economic effect of population aging.
D. Explain an appropriate policy sub-Saharan African countries may implement given their demographic characteristics.
E. Explain one reason why the median age in urban areas of a developed country may be lower than the median age in rural areas.
F. Describe one way super-aged countries such as Japan and Italy can cope with an aging population.
G. Explain one limitation of the data shown in Figure 14.2 in understanding the geographic patterns of aging.

Notes:

Unit 3
Cultural Patterns and Processes

Unit Summary

In Unit 3, you will learn about the most important aspects of culture such as languages, religions, and ethnicities and the powerful role they play in influencing individual and communal behaviors. You'll see that all economic activities and political discussions have a cultural component because culture is a part of everything, and everything is a part of culture.

Modules in Unit 3

What do you see?

Look closely at the unit-opening photograph. Write down 5 things you observe about it. You will return to this image in the AP® Seeing Geography feature at the end of the unit.

1)

2)

3)

4)

5)

Module 18
Introduction to Culture

Before You Read the Module

Focus on Learning Goals

Use the Module learning goals as a way to divide or "chunk" the modules into parts. After you read a section of the module and answer the questions in this book, you should stop and ask yourself: "do I sufficiently understand the goal?"

LG 18-1: Explain how architecture, land use, and food preferences are shaped by culture.
LG 18-2: Describe how cultural traits and attitudes impact place.
LG 18-3: Explain how language, religion, and ethnicity are core parts of cultural identities.
LG 18-4: Define cultural relativism and ethnocentrism.

Key Terms

Take a look at the key terms you will encounter in this module.

long-lot settlement pattern
material culture
non-material culture
cultural trait
local culture
indigenous culture
popular culture
cultural attitudes
language
polyglot
religion
ethnic group
race
ethnic geography
multiculturalism
ethnocentric approach
cultural relativism

While You Read the Module

Answer the questions as you read Module 18: Introduction to Culture.

LG 18-1: ***Explain how architecture, land use, and food preferences are shaped by culture.***

1) Explain how adobe homes in the American Southwest reflect the use of resources of the local environment.

2) Explain how culture influenced the long-lot settlement pattern.

3) Identify the two factors that have historically shaped food preferences.

LG 18-2: ***Describe how cultural traits and attitudes impact place.***

4) Give an example of a cultural trait evident in Figure 18.5 (p. 244).

5) Explain one difference between material and nonmaterial culture.

6) Explain one difference between local and indigenous cultures.

7) Complete the following table.

Cultural Trait	*How Attitude May Vary Between Cultures*
Beliefs about the physical environment	
Family size	
Geographical setting	

LG 18-3: ***Explain how language, religion, and ethnicity are core parts of cultural identities.***

8) Identify the cultural trait that is the main vehicle by which learned belief systems, customs, and skills pass from one generation to the next.

9) Describe how being a polyglot can influence a person's cultural identity.

10) Explain how religion may influence cultural practices.

11) Describe the cultural traits on which the following ethnic groups may base their identity.

Ethnic Group	*Cultural Traits*
Jews	
Amish	
Swiss Americans	
German Americans	
African Americans	

12) Using Figure 18.7, explain how the concept of race shaped colonial Mexico.

LG 18-4: *Define cultural relativism and ethnocentrism.*

13) Explain how most countries have historically promoted a national cultural identity.

14) Explain why some countries have begun to embrace multiculturalism.

15) Imagine an American woman who is traveling throughout East Asia and encounters cultural differences in eating utensils. Explain how she might evaluate those differences.

Attitude toward cultural difference	*Description of attitude toward cultural difference*
Ethnocentric approach	
Cultural relativism	

16) Explain how the international community views human rights when considering issues such as slavery, child marriage, forced marriage, and spousal abuse in the context of attitudes toward cultural differences.

After You Read the Module

Review the Key Terms

Match each key term on the left with its definition on the right.

____ long-lot settlement pattern	1) Intangible elements of culture including a wide range of beliefs, values, myths, and symbolic meanings passed from generation to generation within a given society.
____ material culture	2) A single aspect of a given culture or society.
____ nonmaterial culture	3) Rural, ethnically homogenous culture that is deeply connected to the local land; the opposite of a popular culture.
____ cultural trait	4) The physical, visible objects made and used by members of a cultural group; includes buildings, furniture, clothing, food, artwork, and musical instruments.
____ local culture	5) A local culture that is no longer the dominant ethnic group within its traditional homeland because of migration, colonization, or political marginalization.
____ indigenous culture	6) A linear settlement pattern in which each farmstead is situated at one end of a long, narrow rectangular lot; each lot has access to a major linear resource, usually a river or a major road.
____ popular culture	7) Concepts and ideas in a society that are shaped by cultural opinions, beliefs, and perspectives.
____ cultural attitudes	8) Heterogeneous culture that is more influenced by key urban areas and quick to adopt new technologies; the opposite of a local culture.

____ language	9) A person who is fluent in more than two languages
____ polyglot	10) A people of common ancestry and cultural tradition; characterized by a strong feeling of group identity
____ religion	11) A set of policies that promote the active participation and inclusion of minority groups in national histories, national politics, and cultural institutions with the goal of embracing difference within society
____ ethnic group	12) An approach to understanding other cultures that evaluates them from the perspective of the observer's culture
____ race	13) The study of the spatial aspects of ethnicity
____ ethnic geography	14) An approach to understanding other cultures that seeks to understand individuals and cultures from a wider perspective of cultural logic
____ multiculturalism	15) A mutually agreed-upon system of symbolic communication
____ ethnocentric approach	16) Historically defined by the physical characteristics of a group, especially skin color
____ cultural relativism	17) A structured set of beliefs and practices through which people seek mental and physical harmony with the powers of the universe

Module 19
Cultural Landscapes

Before You Read the Module

Focus on Learning Goals

Use the Module learning goals as a way to divide or "chunk" the modules into parts. After you read a section of the module and answer the questions in this book, you should stop and ask yourself: "do I sufficiently understand the goal?"

> **LG 19-1:** Describe the elements of the cultural landscape.
> **LG 19-2:** Explain how cultural identities are visible in the cultural landscape.
> **LG 19-3:** Explain how attitudes, values, and power relationships shape the cultural landscape.

Key Terms

Take a look at the key terms you will encounter in this module.

physical landscape
placelessness
modernist architecture
postmodern architecture
sequent occupance
sacred spaces
secular
subculture

While You Read the Module

Answer the questions as you read Module 19: Cultural Landscapes.

LG 19-1: ***Describe the elements of the cultural landscape.***

1) Explain how the physical landscape can be reflected in the cultural landscape.

2) Describe how the built environment may contribute to a feeling of placelessness.

3) Identify the land use that is the most dominant feature of the rural landscape.

4) Identify the type of architecture reflected by the Guggenheim Museum in Figure 19.6 (p. 255).

5) Describe how sequent occupance has influenced the cultural landscape of the city of Jerusalem.

LG 19-2: ***Explain how cultural identities are visible in the cultural landscape.***

6) Using Figure 19.9 (p. 257), explain how migration has influenced the ethnic cultural landscape in San Francisco's Mission District.

7) Describe two characteristics of the cultural landscape that can be revealed by publicly displayed linguistic features.

8) Using Figure 19.8 (p. 256), describe how the Dome of the Rock is a feature of the religious landscape in Jerusalem.

LG 19-3: ***Explain how attitudes, values, and power relationships shape the cultural landscape.***

9) Explain how attitudes toward gender can be revealed by the cultural landscape.

10) Explain how the debate over monuments and statues that commemorate the Confederacy illustrates how power relationships shape the cultural landscape.

11) Explain how values related to cleanliness shape the cultural landscape in Japan.

After You Read the Module

Review the Key Terms

Match each key term on the left with its definition on the right.

____ physical landscape	1) A functional, rational, and orderly style for building designs.
____ placelessness	2) Refers to the fact that many places have been controlled or affected by a variety of groups over a period of time; those groups have reshaped the functions or meanings of those places and left behind layers of meaning.
____ modernist architecture	3) Less influenced or controlled by religion.
____ postmodern architecture	4) A group of people with distinct norms, values, and material practices that differentiate them from the dominant culture surrounding them.
____ sequent occupance	5) All the natural physical surroundings that create and shape the places we are living in or examining.
____ sacred spaces	6) A design style that is a reaction against modernist architecture; it has a flair for the dramatic, creating a spectacle while serving a variety of functions.
____ secular	7) Natural or human-made sites that possess religious meaning and are recognized as worthy of devotion, loyalty, fear, or esteem.
____ subculture	8) The feeling resulting from the standardization of the built environment; occurs where local distinctiveness is erased and many places end up with similar cultural landscapes.

Module 20
Cultural Patterns

Before You Read the Module

Focus on Learning Goals

Use the Module learning goals as a way to divide or "chunk" the modules into parts. After you read a section of the module and answer the questions in this book, you should stop and ask yourself: "do I sufficiently understand the goal?"

LG 20-1: Explain how regional cultural patterns contribute to a sense of place.
LG 20-2: Explain how cultural traits can be centripetal forces.
LG 20-3: Explain how cultural traits can be centrifugal forces.

Key Terms

Take a look at the key terms you will encounter in this module.

sense of place
placemaking
centripetal force
centrifugal force
secularization

While You Read the Module

Answer the questions as you read Module 20: Cultural Patterns.

LG 20-1: ***Explain how regional cultural patterns contribute to a sense of place.***

1) Describe how a strong sense of place can be used to develop tourism.

2) Placemaking typically involves efforts at designing public spaces to foster a stronger sense of community. Explain how these efforts can sometimes be contentious.

LG 20-2: ***Explain how cultural traits can be centripetal forces.***

3) Describe how shared cultural traits generally impact the amount of tension and conflict within a society.

4) Complete the following table explaining how particular cultural traits serve as centripetal forces.

Centripetal force	*How cultural trait unifies a neighborhood, society, or country*
Religion	
Language	
Ethnicity	

5) Using Figure 20.2 (p. 267), identify the country in which the ability to speak the native language is perceived as the most important.

LG 20-3: *Explain how cultural traits can be centrifugal forces.*

6) Complete the following table explaining how particular cultural traits serve as centrifugal forces.

Centrifugal force	*How cultural trait threatens the cohesion of a neighborhood, society, or country*
Religion	
Language	
Ethnicity	

7) Using Figure 20.4 (p. 269), explain how linguistic diversity in the Caucasus Mountain region may serve as a centrifugal force.

After You Read the Module

Review the Key Terms

Match each key term on the left with its definition on the right.

____ sense of place	1) Efforts to use and design public places to better serve the needs of residents and to foster a stronger community.
____ placemaking	2) A force that threatens the cohesion of a neighborhood, society, or country.
____ centripetal force	3) The process whereby religion becomes a less dominant force in everyday life than it was in the past.
____ centrifugal force	4) The distinctive feeling of a place, or a person's perception of place.
____ secularization	5) A force that brings people together and unifies a neighborhood, society, or country.

Module 21

Types of Diffusion and Their Historical Causes

Before You Read the Module

Focus on Learning Goals

Use the Module learning goals as a way to divide or "chunk" the modules into parts. After you read a section of the module and answer the questions in this book, you should stop and ask yourself: "do I sufficiently understand the goal?"

LG 21-1: Describe the types of diffusion.

LG 21-2: Explain how global forces can create new forms of expression.

LG 21-3: Explain how colonialism, imperialism, and trade shaped cultural patterns and practices.

Key Terms

Take a look at the key terms you will encounter in this module.

absorbing barriers
permeable barriers
pidgin
creole
creolization
lingua franca
bilingualism
empire
imperialism
colonialism
genocide

While You Read the Module

Answer the questions as you read Module 21: Types of Diffusion and Their Historical Causes.

LG 21-1: ***Describe the types of diffusion.***

1) Describe the two main types of diffusion.

2) Complete the following table with examples of specific types of expansion diffusion.

Type of Expansion Diffusion	*Example*
Hierarchical	
Contagious	
Stimulus	

LG 21-2: ***Explain how global forces can create new forms of expression.***

3) Explain how trade influences the development of a pidgin language.

4) Describe the difference between a pidgin and a creole language.

5) Explain why English is considered a global lingua franca.

LG 21-3: ***Explain how colonialism, imperialism, and trade shaped cultural patterns and practices.***

6) Describe a specific way in which colonialism and imperialism influenced cultural patterns in a place that was colonized.

7) Using Figure 21.4 (p. 278), explain how colonial and imperialism can be seen in the cultural landscape.

8) Explain how colonialism and imperialism influenced the distribution of Islam in South America.

After You Read the Module

Review the Key Terms

Match each key term on the left with its definition on the right.

____ absorbing barriers	1) Barriers that completely halt diffusion.
____ permeable barriers	2) Barriers that slow diffusion, but still allow some partial or weakened diffusion.
____ pidgin	3) The ability to speak two languages fluently.
____ creole	4) The systematic killing of members of a racial, ethnic, or linguistic group.
____ creolization	5) The act of forcefully controlling a foreign territory, which becomes known as a colony.
____ lingua franca	6) A combined language that has a fuller vocabulary than a pidgin language and becomes a native language.
____ bilingualism	7) The linguistic process where languages converge and create new languages and forms of communication.
____ empire	8) A sovereign political entity that seeks to expand beyond their origin land to control more territory politically and/or economically.
____ imperialism	9) A trade language, characterized by a very small vocabulary derived from the languages of at least two or more groups in contact.
____ colonialism	10) The motivating impulse to control greater amounts of territory.
____ genocide	11) A language of communication and commerce spoken across a wide area where it is not a mother tongue.

Module 22
Contemporary Causes of Diffusion

Before You Read the Module

Focus on Learning Goals

Use the Module learning goals as a way to divide or "chunk" the modules into parts. After you read a section of the module and answer the questions in this book, you should stop and ask yourself: "do I sufficiently understand the goal?"

LG 22-1: Describe how modern forces are reshaping cultural patterns.

LG 22-2: Explain how communication technologies impact human interactions.

Key Terms

Take a look at the key terms you will encounter in this module.

time-space convergence
endangered language
extinct language
convergence hypothesis
glocalization

While You Read the Module

Answer the questions as you read Module 22: Contemporary Causes of Diffusion.

LG 22-1: ***Describe how modern forces are reshaping cultural patterns.***

1) Explain how urbanization can influence popular culture.

2) Complete the following table.

Effects of Globalization on Cultures of Indigenous peoples	*Description*
Cultural survival	
Cultural endangerment	

3) Describe how the experience of people having Japanese ancestry who were born in Brazil illustrates pressures faced by ethnic minorities.

LG 22-2: *Explain how communication technologies impact human interactions.*

4) Describe spatial patterns in Google searches about Jesus compared to searches about Allah in Figure 22.4 (p. 286).

5) Describe how advances in transportation have facilitated time-space convergence.

6) Using Figure 22.6 (p. 288), identify the second most prevalent language used on the Internet.

7) Explain how migration has impacted the linguistic diversity of New York City.

8) Describe how communication technologies can help preserve endangered languages.

9) Complete the following table.

Effects of Globalization on Culture	*Description*
Cultural convergence	
Cultural divergence	

After You Read the Module

Review the Key Terms

Match each key term on the left with its definition on the right.

____ time-space convergence	1) A language that has only a few elderly speakers still living or no living speakers.
____ endangered language	2) Adapting global practices to fit local cultural practices and preferences.
____ extinct language	3) The phenomenon whereby the introduction of new transportation technologies progressively reduces the time it takes to travel between places.
____ convergence hypothesis	4) The idea that cultures are converging, or becoming more alike.
____ glocalization	5) A language that is not taught to children by their parents and is not used actively in everyday matters.

Module 23
Diffusion of Language and Religion

Before You Read the Module

Focus on Learning Goals

Use the Module learning goals as a way to divide or "chunk" the modules into parts. After you read a section of the module and answer the questions in this book, you should stop and ask yourself: "do I sufficiently understand the goal?"

LG 23-1: Describe how cultural systems form.
LG 23-2: Describe how cultural systems diffuse.
LG 23-3: Explain how maps and charts can show diffusion patterns.
LG 23-4: Explain how universalizing religions spread.
LG 23-5: Describe the spatial patterns of ethnic religions.

Key Terms

Take a look at the key terms you will encounter in this module.

culture hearth
language family
dialect
accent
monotheistic
universalizing religion
ethnic religion
proselytic
toponyms
generic toponym
polytheistic
animistic religion

While You Read the Module

Answer the questions as you read Module 23: Diffusion of Language and Religion.

LG 23-1: ***Describe how cultural systems form.***

1) Explain why selected elements of a culture, such as holidays and festivals, are often the parts of ethnicity that survive when a culture diffuses far from its hearth.

2) Explain how imperialism and colonialism diffused Indo-European languages far from their hearths.

3) Use Figure 23.1 (p. 294-295) to compare the distribution of the Indo-European language family and the Sino-Tibetan language family.

4) Complete the following table using Figure 23.5 (p. 297).

Region of Religious Hearth	*Religion(s) and regions to which each diffused*
Middle East	
South Asia	
East Asia	

LG 23-2: ***Describe how cultural systems diffuse.***

5) Identify the three major world religions that arose among people who spoke Semitic languages in the Middle East.

6) Explain why universalizing religions like Christianity and Islam are more likely to diffuse than ethnic religions such as Judaism.

7) Explain why Hinduism remains clustered near its hearth.

8) Explain how Confucianism and Taoism were spread through hierarchical diffusion.

LG 23-2: ***Explain how maps and charts can show diffusion patterns.***

9) Describe how Indo-European languages diffused through colonialism and imperialism.

10) Identify the language family that has the second highest number of native speakers.

11) Identify the Semitic language having the highest number of native speakers.

12) Describe how toponyms become part of the cultural landscape.

13) Explain why the government of India began referring to the country as Bharat in 2023.

LG 23-4: ***Explain how universalizing religions spread.***

14) Identify the two universalizing religions having the greatest number of people who practice them.

15) Identify the division of Islam to which the majority of Muslims belong.

16) Identify the country having the world's largest Islamic population in terms of numbers.

17) Explain why Islam is the world's fastest-growing religion.

18) Explain how Buddhism diffused from its hearth.

LG 23-5: ***Describe the spatial patterns of ethnic religions.***

19) Describe one difference between universalizing and ethnic religions.

20) Explain why Hindus are more likely than adherents to any other major world faith to constitute a majority in the countries in which they reside.

21) Explain why migration plays a key role in the diffusion of ethnic religions.

After You Read the Module

Review the Key Terms

Match each key term on the left with its definition on the right.

____ culture hearth	1) Relating to the belief in only one god.
____ language family	2) A regional variation of a language that is understood by people who speak other variations of that language.
____ dialect	3) A group of related languages that share a common ancestry.
____ accent	4) A way of pronouncing words.
____ monotheistic	5) A focused geographic area where important innovations are born and from where they spread.
____ universalizing religion	6) The generic part of a place-name, often a suffix or prefix, such as -ville in Louisville.
____ ethnic religion	7) A faith that subscribes to the idea that souls or spirits exist not only in humans, but also in animals, plants, rocks, natural phenomena such as thunder, geographic features such as mountains or rivers, and other entities of the natural environment.

____ proselytic	8) The names given to places
____ toponyms	9) A religion that actively seeks new members and believes its message has universal importance and application
____ generic toponym	10) Describing a religion that spreads its message to others through missionary work
____ polytheistic	11) A religion identified with a particular ethnic or tribal group that does not seek converts
____ animistic religion	12) Relating to the belief in many gods

Module 24
Effects of Diffusion

Before You Read the Module

Focus on Learning Goals

Use the Module learning goals as a way to divide or "chunk" the modules into parts. After you read a section of the module and answer the questions in this book, you should stop and ask yourself: "do I sufficiently understand the goal?"

LG 24-1: Describe the effects of cultural diffusion on peoples and places.
LG 24-2: Explain how syncretism is a result of cultural diffusion.

Key Terms

Take a look at the key terms you will encounter in this module.

acculturation
assimilation
transculturation
syncretism
syncretic religion
orthodox religion

While You Read the Module

Answer the questions as you read Module 24: Effects of Diffusion.

LG 24-1: ***Describe the effects of cultural diffusion on peoples and places.***

1) Describe an example of how an element of popular culture has diffused widely.

2) Explain why acculturation is frequently seen in the adoption of a new language by first-generation immigrants.

3) Explain how assimilation can be seen in the example of German Americans.

4) Explain why multiculturalism is increasingly a key goal in many places.

LG 24-2: ***Explain how syncretism is a result of cultural diffusion.***

5) Identify the cultural trait typically associated with syncretism.

6) Explain why many religions have orthodox strains.

7) Explain why Sikhism is considered by many to be the best example of religion syncretism on a grand scale.

8) Describe how forces of globalization are likely to impact the practice of religion.

After You Read the Module

Review the Key Terms

Match each key term on the left with its definition on the right.

____ acculturation	1) Occurs when an ethnic or immigrant group blends in with the host culture and loses many culturally distinctive traits.
____ assimilation	2) Religion that combines elements of two or more different belief systems.
____ transculturation	3) Religion that emphasizes purity of faith and is generally not open to blending with elements of other belief systems.
____ syncretism	4) Occurs when an ethnic or immigrant group adopts enough of the ways of the host society to be able to function economically and socially.
____ syncretic religion	5) The notion that people adopt elements of other cultures as well as contribute elements of their own culture, thereby transforming both cultures.
____ orthodox religion	6) The blending of beliefs, ideas, practices, and traits, especially in a religious context.

Unit 3
Review Exercises

Check Your Understanding

Review the "Module Review" summaries at the end of each module in Unit 3 of your textbook. Make sure you understand all of the concepts before moving on to the next unit.

Unit 3 Multiple-Choice Question AP® Exam Practice

1. Which cultural trait is most evident in Figure 18.2 (p. 242)?
   (A) Architecture
   (B) Language
   (C) Religion
   (D) Land use
   (E) Food preferences

2. Sites in Jerusalem in Figure 19.8 (p. 256) associated with Islam and Judaism best reflect the concept of
   (A) Language
   (B) Placelessness
   (C) Postmodern architecture
   (D) Indigenous ethnicity
   (E) Sequent occupance

3. Using Figure 20.4 (p. 269), the Kurdish language is most likely to function in Armenia as which of the following?
   (A) A centripetal force at the national scale
   (B) A centrifugal force at the national scale
   (C) A centripetal force at the local scale
   (D) Both a centrifugal and centripetal force at the national scale
   (E) Both a centrifugal and centripetal force at the local scale

4. The type of diffusion in which a cultural trait spreads with the movement of people is
   (A) Expansion
   (B) Relocation
   (C) Stimulus
   (D) Contagious
   (E) Hierarchical

5. In communities along the U.S.-Mexico border, residents have developed unique linguistic patterns that combine recognizable elements of both English and Spanish. It does not have a full-fledged vocabulary with words not spoken in any other language and is not spoken anyone as a native language. This best represents
   (A) The process of creolization.
   (B) The failure of proper language education in both countries.
   (C) The inevitable loss of traditional linguistic purity.
   (D) A new form of cultural expression that serves their unique communication needs.
   (E) The eventual dominance of one language over the other in border regions.

6. Time-space convergence has impacted global cultural interactions most directly by
   (A) Completely eliminating all cultural differences between groups across the world.
   (B) Dramatically increasing the speed and intensity of cultural exchanges due to advances in technology.
   (C) Making traditional cultural practices completely obsolete in the modern world.
   (D) Leading to complete homogenization of global cultural practices.
   (E) Making cultural differences more pronounced due to increased awareness of global diversity.

7. Using Figure 23.14 (p. 305), which of the following best explains a strength of the map at the local scale of analysis as compared to the national scale of analysis?
   (A) The map at the local scale of analysis shows that most people who are practicing Southern Baptists live in the western United States.
   (B) The map at the local scale of analysis shows the significant number of people who are members of the Church of Jesus Christ of Latter-day Saints.
   (C) The map at the local scale of analysis shows that people who are members of the Evangelical Lutheran Church in America are clustered in the southern United States.
   (D) The map at the local scale of analysis shows that the Catholic population is clustered only in urban areas.
   (E) The map at the local scale of analysis shows the spatial patterns associated with various branches and denominations of Christianity.

8. In the Kensington Market neighborhood of Toronto, Canada, Orthodox Jewish synagogues are next to Buddhist temples, and halal butcher shops are near Italian delis. Signs are printed in multiple languages, and clothing associated with traditional instead of Western or popular culture is worn by some residents. This local area best represents which of the following?
   (A) Cultural syncretism in an urban area
   (B) Multiculturalism resulting from cultural diffusion
   (C) The process of cultural assimilation
   (D) Neocolonialism in North America
   (E) The failure of immigrants to assimilate

Unit 3 Free-Response Question AP® Exam Practice

Indigenous cultures are a specific type of local cultures. According to the United Nations, 6.2% of the world's population is Indigenous. They live in 90 countries and in more than 5000 distinct groups with distinct identities and cultures. Respond to parts A, B, C, D, E, F, and G.

A. Describe the concept of a local culture.
B. Explain why Indigenous cultures are no longer the dominant ethnic group within their traditional homelands.
C. Explain how Indigenous culture may play a role in placemaking.
D. Explain how economic globalization can threaten Indigenous lands.
E. Describe how Indigenous cultures can use communication technologies to facilitate cultural survival.
F. Explain why syncretic religions in the Americas often incorporated elements of Indigenous religious beliefs.
G. Explain the degree to which many Indigenous languages are likely to be considered endangered.

Notes:

Unit 4
Political Patterns and Processes

Unit Summary
Political geography is a branch of human geography concerned with the spatial analysis of political systems. We will examine the creation and impact of political boundaries, the development of states and nations, and the relationships between sovereignty, governance, and power. Ultimately, we will analyze how political power has been and continues to be distributed and contested across all scales. We will look at the impact of colonialism and imperialism historically and how supranationalism and devolution impact states in the contemporary world. The unit also covers how boundaries are established and contested—on land and at sea—and how geography influences political conflicts, cooperation, and territorial disputes.

Modules in Unit 4
Module 25: Introduction to Political Geography
Module 26: Political Processes
Module 27: Political Power and Territoriality
Module 28: Types and Functions of Political Boundaries
Module 29: Internal Political Boundaries
Module 30: Forms of Governance
Module 31: Challenges to State Sovereignty: Devolution and Supranationalism
Module 32: Consequences of Centrifugal and Centripetal Forces

What do you see?
Look closely at the unit-opening photograph. Write down 5 things you observe about it. You will return to this image in the AP® Seeing Geography feature at the end of the unit.

1)

2)

3)

4)

5)

Module 25
Introduction to Political Geography

Before You Read the Module

Focus on Learning Goals

Use the Module learning goals as a way to divide or "chunk" the modules into parts. After you read a section of the module and answer the questions in this book, you should stop and ask yourself: "do I sufficiently understand the goal?"

LG 25-1 Describe the world political map and what we can learn from it.

LG 25-2 Define the different types of political entities.

Key Terms

Take a look at the key terms you will encounter in this module.

political geography
political map
state or country
independent state
sovereign state
nation
nation-state
nation-state ideal
nationalism
stateless nation
multistate nations
autonomous region

While You Read the Module

Answer the questions as you read Module 25: Introduction to Political Geography.

LG 25-1: ***What is the world political map, and what can we learn from it?***

1) Define the term "state" as used in a political geographic unit and identify a term that is similar that is mentioned in the reading.

2) States are the building blocks in today's world, but come in different shapes, sizes, and populations. Use Figure 25.1 (p. 338-339) and identify 5 large states in area and 5 small states.

3) Define sovereignty as it applies to independent states.

LG 25-2: ***What are the different types of political entities?***

4) There are many different terms that can be combined to create very specific political entities. Two terms are "nation" and "state". Define these terms and the combinations to understand the ways these terms are used to explain political entities.

Term	*Definition*
Nation	
State	
Nation-State	
Nation State Ideal	
Stateless Nation	

5) Use Figure 25.4 (p. 342) to explain why South Africa is a multinational state.

6) Define *multistate nation.*

7) Describe how autonomous or semiautonomous regions are not independent states, but still have some autonomy.

After You Read the Module

Review the Key Terms Part 1

Match each key term on the left with its definition on the right.

____ political geography	1) A community of people bound to a homeland and possessing a common identity.
____ political map	2) A state that possesses the sole authority over the land and the people within its boundaries.
____ sovereign state	3) A branch of human geography, concerned with the spatial analysis of political systems.
____ nation	4) The ideal political geographic unit is where a nation of people is within a state's territorial boundaries.
____ nation-state	5) A map that shows the spatial organization of the countries and territories of the entire globe at a given point in time.
____ nation-state ideal	6) The idea that political authorities govern in the name of all the country citizens, modern mass media link all residents, and state-based citizenship, rights reinforced the idea of a national identity.

Review the Key Terms Part 2

Match each key term on the left with its definition on the right.

____ stateless nation	1) A country containing multiple National, ethnic, and religious groups within its boundaries.
____ multinational state	2) Ethnic groups, territories divided by one or more international boundaries.
____ multistate nation	3) An ethnic group or nation that does not possess its own state and is not the majority population in any state.
____ autonomous region	4) A subdivision or dependent territory of a country that has a degree of self-government in its decision-making.

Module 26

Political Processes

Before You Read the Module

Focus on Learning Goals

Use the Module learning goals as a way to divide or "chunk" the modules into parts. After you read a section of the module and answer the questions in this book, you should stop and ask yourself: "do I sufficiently understand the goal?"

LG 26-1 Identify the concepts that shape the political geography of today's world.

LG 26-2 Explain the processes that influence political boundaries.

Key Terms

Take a look at the key terms you will encounter in this module.

self-determination
core area
ethnonationalism
devolution

While You Read the Module

Answer the questions as you read Module 26: Political Processes.

LG 26-1: ***What concepts shape the political geography of today's world?***

1) Define *self-determination* and explain how it applies to the American Revolution.

2) Core areas are a small territorial nucleus from which a country grows in area and over time. Describe the physical geographic features often associated with the development and growth of a core area.

3) Compare France and the Democratic Republic of the Congo's development of a core area and how it impacts conflicts within a state (country).

4) Explain the degree that which globalization has impacted the power of states according to economists.

LG 26-2: *What processes influence political boundaries?*

5) Using Figure 26.3 (p. 351), identify the two major colonial powers in Africa.

6) Using Figure 26.3 (p. 351), compare European colonial boundaries with the ethnic boundaries within Africa.

7) Using Figure 26.4 (p. 352), identify the two major colonial powers in Latin America.

8) The textbook states that "Colonialism is the result of imperialism" twice in the reading. Describe what this means.

9) Describe what occurred with African civil servants that led them to start nationalist movements to oppose colonial rule.

10) Using Figure 26.5 (p. 353), identify the period when most African countries achieved independence.

11) Using Figure 26.6 (p. 354), compare the independence time frames with those from Africa.

12) Define *devolution*.

13) Explain how ethnonationalism can be a devolutionary force.

14) Explain how economic forces can lead to devolution.

15) Explain how spatial or territorial forces can lead to devolution.

After You Read the Module

Review the Key Terms

Match each key term on the left with its definition on the right.

____ self-determination	1) The movement of power from a central government to a regional government within a state.
____ core area	2) A form of nationalism in which a nation is defined in terms of ethnic identity.
____devolution	3) A nation's ability to determine its statehood and form its allegiance and government.
____ ethnonationalism	4) A small territorial nucleus from which a country grows in area and over time.

Complete the following chart to make sure you understand these key concepts.

Term	*Definition and example*
Colonization	
Imperialism	
Independence movements	

Module 27

Political Power and Territoriality

Before You Read the Module

Focus on Learning Goals

Use the Module learning goals as a way to divide or "chunk" the modules into parts. After you read a section of the module and answer the questions in this book, you should stop and ask yourself: "do I sufficiently understand the goal?"

LG 27-1 Identify the concepts geographers use to describe political power.
LG 27-2 Describe territoriality in political geography.

Key Terms

Take a look at the key terms you will encounter in this module.

neo-colonialism	boundary
peripheral states	median line principle
core states	borderland
shatterbelts	frontier
choke point	enclave
strait	exclave

While You Read the Module

Answer the questions as you read Module 27: Political Power and Territoriality.

LG 27-1: ***What concepts do geographers use to describe political power?***

1) Define *neo-colonialism* and describe how it is different from colonialism.

2) Compare core states to peripheral states.

3) Using the example of a cocoa farmer in West Africa (p. 363-364) and the idea of dependence of a single product, explain how peripheral states are dependent on core countries for income.

4) Define *shatterbelt* and explain why Eastern Europe is considered a shatterbelt region.

5) Define *choke point* and explain how the Strait of Hormuz is a choke point.

6) Explain how political stability and control can impact choke points economically.

LG 27-2: *What is territoriality in political geography?*

7) Define *territoriality*.

8) Explain how territoriality and independence movements presented a clash between Russia and Ukraine.

9) Define *boundary*.

10) Explain how "Tex-Mex" culture is captured in the idea of a borderland.

After You Read the Module

Review the Key Terms

Match each key term on the left with its definition on the right.

____ neo-colonialism	1) A territory surrounded by a country, but not ruled by it.
____ peripheral states	2) A region straddling both sides of an international boundary.
____ core states	3) A narrow body of water connecting to larger bodies.
____ shatterbelt	4) A narrow passage that restricts traffic to another region.
____ choke point	5) A Region of continuing and persistent fragmentation due to devolution and centrifugal forces.
____ strait	6) States that have relatively little industrial development and focus mostly on agriculture and raw materials.
____ boundary	7) The set of economic and political strategies by which wealthier and powerful countries indirectly maintain influence over less wealthy areas.
____ median line principle	8) States that have the most advanced industrial and military technologies and the highest levels of wealth and mass consumption.
____ frontier	9) A demarked line that marks both the limits of territory and divisions between territories.
____ enclave	10) An approach to dividing and creating boundaries at the midpoint between two places.

Module 28

Types and Functions of Political Boundaries

Before You Read the Module

Focus on Learning Goals

Use the Module learning goals as a way to divide or "chunk" the modules into parts. After you read a section of the module and answer the questions in this book, you should stop and ask yourself: "do I sufficiently understand the goal?"

LG 28-1 Describe the different types of country boundaries.
LG 28-2 Describe the characteristics of special boundaries.
LG 28-3 Explain how boundaries influence identities and interactions between governments.
LG 28-4 Explain how the United Nations Convention on the Law of the Sea affects international relations.

Key Terms

Take a look at the key terms you will encounter in this module.

delimited
demarcated
relic boundary
superimposed boundary
subsequent boundary
antecedent boundary
geometric boundary
consequence boundary
demilitarized zone (DMZ)
buffer state
satellite state
United Nations Convention on the Law of the Sea
Exclusive Economic Zone
Arctic Council

While You Read the Module

Answer the questions as you read Module 28: Types and Functions of Political Boundaries.

LG 28-1: ***What are the different types of country boundaries?***

1) Using Figure 21.1A (p.3 80), describe how Hadrian's Wall is an example of a demarcated boundary.

2) Using Figure 21.1B (p. 380), explain why Hadrian's Wall is considered a relic boundary today.

3) Complete the table below.

Type of Boundary	*Definition*	*Example*
Superimposed Boundary		
Subsequent Boundary		
Antecedent Boundary		
Geometric Boundary		
Consequent Boundary		

LG 28-2: ***What are the characteristics of special boundaries?***

4) Using Figure 28.9 (p. 386), describe why the boundary between North and South Korea is defined as a demilitarized zone.

5) Using Figure 28.11 (p. 388), describe how the boundaries established at the Berlin Conference would be considered superimposed boundaries.

LG 28-3: ***How do boundaries influence identities and interactions between governments?***

6) Using Figure 28.12 (p. 389), describe how Siam (now Thailand) acted as a buffer state around 1900.

7) Describe how Eastern European states, like Poland, were considered satellite states of the Soviet Union after World War II.

LG 28-4: ***How does the United Nations Convention on the Law of the Sea affect international relations?***

8) Define the United Nations Convention on the Law of the Sea (UNCLOS).

9) Describe what a coastal state has the sole right to do in the exclusive economic zone (EEZ).

10) Use Table 28.1 (p.390) to complete the table below.

Zone	*Distance from the Shoreline*	*Description*
Territorial Sea		
Contiguous Zone		
Exclusive Economic Zone		
International Waters (High Seas)		

11) Explain a limitation of the United Nations Convention on the Law of the Sea (UNCLOS) with regard to the dispute over the Spratly Islands.

12) Describe how global warming will impact the waters in the Arctic Circle.

After You Read the Module

Review the Key Terms

Match each key term on the left with its definition on the right.

____ delimit	1) A boundary that is drawn to accommodate existing cultural differences.
____ demarcated	2) A boundary that is drawn after the cultural landscape is developed.
____ relic boundary	3) A boundary that no longer functions as an international border.
____ superimposed boundary	4) Describing how boundaries are fixed or defined to identify their limits.
____ subsequent boundary	5) A boundary that is placed on an area without regard to existing boundaries.
____ antecedent boundary	6) A boundary that was identified before an area was settled.
____ geometric boundary	7) Describing how boundaries are set apart to distinguish their limits.
____ consequent boundary	8) A boundary that has regular, often perfect straight lines, drawn without regard for the areas, physical or cultural features.

Review the Key Terms Part 2

Match each key term on the left with its definition on the right.

____ buffer state	1) The area that extends 200 nautical miles from the shoreline in which coastal states have the sole right to exploit, develop, manage, and conserve all water resources lying beyond the land.
____ satellite state	2) A nominally independent country that is politically, militarily, and economically controlled by a more powerful state.
____ United Nations Convention on the Law of the Sea (UNCLOS)	3) Area defined by the 66° 34 minute north latitude line.
____ exclusive economic zone (EEZ)	4) A politically and economically weak, independent country that lies between the borders of two powers.
____ Arctic Circle	5) Conference organized to define the territorial boundaries and the right to the sea.

Module 29
Internal Political Boundaries

Before You Read the Module

Focus on Learning Goals

Use the Module learning goals as a way to divide or "chunk" the modules into parts. After you read a section of the module and answer the questions in this book, you should stop and ask yourself: "do I sufficiently understand the goal?"

LG 29-1 Explain how geography and internal political boundaries can shape voting outcomes.
LG 29-2 Explain how voting outcomes can reflect regional differences.

Key Terms

Take a look at the key terms you will encounter in this module.

electoral geography	redistricting
voting district	gerrymandering
electoral college	packing
reapportionment	cracking

While You Read the Module

Answer the questions as you read Module 29: Internal Political Boundaries.

LG 29-1: ***How can geography and internal political boundaries shape voting outcomes?***

1) Define *electoral geography*.

2) Voting districts happen at various scales. Using Purple Sage, Wyoming as an example, describe how voters are in several different-sized districts.

3) Define *electoral college*.

4) Using Figure 29.1 (p. 400), explain why New Jersey currently has 14 delegates while Wisconsin, a larger state, has only 10 delegates.

5) There are 435 seats in the House of Representatives. Explain why reapportionment occurs after every census.

6) Using Figure 29.3 (p. 401), identify which region(s) lost seats after the 2020 Census.

7) Complete the following chart.

Term	*Definition*
Redistricting	
Gerrymandering	
Packing	
Cracking	

8) Use Figure 29.5 (p. 403), explain how gerrymandering can turn a state with a 60% "purple" districts into a majority "gold" state.

9) Explain why the map of Wisconsin's congressional districts changed so much from 2008-2024.

LG 29-2: ***How can voting outcomes reflect regional differences?***

10) Using Figures 29.8 (p. 405) and 29.9 (p. 406), explain why a cartogram might be a better map to communicate who won the 2020 presidential election.

11) Using Figure 29.10 (p. 406), explain how changing the scale of analysis to counties impacts our understanding of voting patterns in the United States.

12) Using Figure 29.11 (p. 407), explain how the three-dimensional cartogram helps give a sense of voting patterns in urban and rural areas.

After You Read the Module

Review the Key Terms

Match each key term on the left with its definition on the right.

____ electoral Geography	1) The manipulation of voting district boundaries to favor a political party, group, or election outcome.
____ voting district	2) A body of 538 electors in the United States.
____ electoral college	3) A subfield of political geography that analyzes the geography of political preferences and how geography can shape voting outcomes.
____ reapportionment	4) The process by which the 435 seats in the House of Representatives are divided proportionately by population among the 50 states.
____ redistricting	5) The two different methods used to gerrymander districts to gain a political advantage.
____ gerrymandering	6) The process of drawing new boundaries for U.S. congressional districts to reflect the population changes since the previous U.S. census.
____ packing/cracking	7) A territorial division for casting votes in public elections.

Module 30
Forms of Governance

Before You Read the Module

Focus on Learning Goals

Use the Module learning goals as a way to divide or "chunk" the modules into parts. After you read a section of the module and answer the questions in this book, you should stop and ask yourself: "do I sufficiently understand the goal?"

LG 30-1 Explain the differences between unitary and federal states.
LG 30-2 Explain the spatial organization of unitary and federal states.

Key Terms

Take a look at the key terms you will encounter in this module.

subnational units
unitary state
federal state

While You Read the Module

Answer the questions as you read Module 30: Forms of Governance.

LG 30-1: ***What are the differences between unitary and federal states?***

1) Define *unitary state.*

2) Describe how a federal state divides power among subnational units.

3) According to the textbook, identify two qualities where a unitary system works best.

4) According to the textbook, identify one quality where a federal system works best.

LG 30-2: ***What are the differences between unitary and federal states?***

5) Explain how Rwanda used the concept of a unitary state to create stability within the country.

6) Explain how federalism helps Canada counteract French Canadian demands for Quebec's independence.

7) Describe how Native American reservations act as autonomous regions outside of the typical federal system.

8) Compare the Canadian Nunavut territory to Native American reservations.

After You Read the Module

Review the Key Terms

Match each key term on the left with its definition on the right.

____ subnational units	1) An independent country that disperses significant authority among subnational units.
____ federal states	2) The smaller areas into which a larger state is divided.
____ unitary states	3) An independent state that concentrates power in the central government and grants little or no authority to it subnational units.

Module 31
Challenges to State Sovereignty: Devolution and Supranationalism

Before You Read the Module

Focus on Learning Goals

Use the Module learning goals as a way to divide or "chunk" the modules into parts. After you read a section of the module and answer the questions in this book, you should stop and ask yourself: "do I sufficiently understand the goal?"

LG 31-1: Identify the devolutionary forces that can challenge state sovereignty.
LG 31-2: Explain how devolutionary challenges occur at different scales.
LG 31-3: Explain how innovations in communication technology challenge sovereignty.
LG 31-4: Identify why supranational organizations are formed and how they challenge state Sovereignty.

Key Terms

Take a look at the key terms you will encounter in this module.

irredentism
terrorism
international terrorism
domestic terrorism
state terrorism
subnational terrorism
ETA
democratization
supranationalism
supranational organization
European Union
African Union
Arctic Council
regional trading bloc
economies of scale
North American Free Trade Agreement (NAFTA-USCMA)
Association of Southeast Asian Nations (ASEAN)
North Atlantic Trade Organization (NATO)

While You Read the Module

Answer the questions as you read Module 31: Challenges to State Sovereignty: Devolution and Supranationalism.

LG 31-1: ***Which devolutionary forces can challenge state sovereignty?***

1) Describe the ethnic cleansing that is occurring in western Myanmar.

2) Complete the chart below regarding different forms of terrorism.

Term	*Definition*	*Example*
International Terrorism		
Domestic Terrorism		
State Terrorism		
Subnational Terrorism		

3) Define irredentism and give one example from the textbook.

LG 31-2: ***How do devolutionary challenges occur at different scales?***

4) Complete the following chart describing why these three new states were created.

Term	*Describe why it disintegrated from another state*
South Sudan	
Eritrea	
Timor-Leste	

LG 31-3: ***How do innovations in communication technology challenge sovereignty?***

5) Describe how the internet and social media challenge the power of state sovereignty.

6) Explain why the U.S. government expressed concern and looked to ban TikTok.

LG 31-4: ***Why are supranational organizations formed, and how do they challenge state sovereignty?***

7) Define *supranationalism*.

8) Identify 3 reasons why states might join a supranational organization.

9) Complete the chart below.

Supranational Organization	*Goals and Purpose of the Organization*
United Nations	
The European Union	
The African Union	
Arctic Council	

10) Using one of the organizations in the chart in question 9, explain how that supranational organization challenges the power of state sovereignty.

11) Describe the purpose of a regional trade bloc.

12) Describe one of the goals of the Association of Southeast Asian Nations (ASEAN).

13) Use Figure 31.14 (p. 439) to describe the spatial distribution of the member states of NATO.

After You Read the Module

Review the Key Terms Part 1

Match each key term on the left with its definition on the right.

____ irredentism	1) A Basque separatist organization in Spain that uses terrorism in its campaign for an independent Basque state.
____ terrorism	2) Acts by individuals or groups against the citizens or government of their own country.
____ international terrorism	3) The calculated use of violent acts against civilians and symbolic targets to publicize a cause, intimidate a civilian population, or affect the conduct of a government.
____ domestic terrorism	4) Terrorism committed by government agents, whose leaders have ordered them to murder, imprison, or force into exile perceived enemies of the state.
____ state terrorism	5) Terrorism that transcends national boundaries and is intended to intimidate people in other countries.
____ subnational terrorism	6) The political claim to territory in another country based on ethnic affiliations and historic borders.
____ ETA	7) Terrorism is committed by non-governmental groups that feel wronged by their government.

Review the Key Terms Part 2

Match each key term on the left with its definition on the right.

____ democratization	1) Cost advantages that can come with a larger scale of operations.
____ supranationalism	2) An international government forum that promotes interaction among Arctic states and indigenous communities on common Arctic issues.
____ African Union	3) A political, economic, and social union of 27 independent European countries that promotes the free movement of people, goods, services, and capital among its members.
____ European Union	4) Occurs when a sovereign state moves from a non-democracy to a democracy.
____ Arctic Council	5) A 1994 trade agreement between Canada, the United States, and Mexico.
____ Free Trade Agreement (FTA)	6) Occurs when a collection of nation states and their citizens relinquish some sovereign rights to a large-scale body that exercises authority over its member states.
____ economies of scale	7) A regional Intergovernmental organization comprising 10 countries in Southeast Asia to promote intergovernmental cooperation and facilitate economic growth, social progress, and cultural development in the region.
____ North American Free Trade Agreement (NAFTA)	8) A multi-country agreement that reduces or eliminates taxes to promote the free flow of goods and services across international boundaries.
____ Association of Southeast Asian Nations (ASEAN)	9) a continental union of 55 member states, aiming to promote unity, solidarity, and development within the African continent.

Module 32
Consequences of Centrifugal and Centripetal Forces

Before You Read the Module

Focus on Learning Goals

Use the Module learning goals as a way to divide or "chunk" the modules into parts. After you read a section of the module and answer the questions in this book, you should stop and ask yourself: "do I sufficiently understand the goal?"

LG 32-1: *Explain centrifugal forces and their impacts on states.*
LG 32-2: *Explain centripetal forces and their impacts on states.*

Key Terms

Take a look at the key terms you will encounter in this module.

failed state
uneven development
allegiance
equitable infrastructure
social cohesion
iconography

While You Read the Module

Answer the questions as you read Module 32: Consequences of Centrifugal and Centripetal Forces.

LG 32-1: ***What are centrifugal forces, and how do they impact states?***

1) Define *failed state*.

2) Describe why Somalia is considered to be a failed state.

3) Using Figure 32.4 (p. 446), describe the pattern of development in the photo.

4) Describe why uneven development is considered a centrifugal force.

5) Describe how stateless nations are considered a centrifugal force.

LG 32-2: ***What are centripetal forces, and how do they impact states?***

6) Explain how nationalism operates as a centripetal force in multi-ethnic countries such as the United States.

7) Using Figure 32.8 (p. 449), explain how the pattern of roads would be a centripetal force in Britain from 43-180 C.E.

8) Explain how the use of iconography would act as a centripetal force for a state.

After You Read the Module

Review the Key Terms

Match each key term on the left with its definition on the right.

____ failed state	1) Cultural unity occurs when the members of a society are culturally united.
____ uneven development	2) Loyalty or commitment to a country.
____ allegiance	3) Occurs when core states have advanced economies and a high standard of living while peripheral states have relatively little industrial development.
____ equitable infrastructure	4) A state whose political or economic system has become so weak that the government is no longer in control
____ cultural cohesion	5) A set of traditional symbols or symbolic forms associated with a country and citizens.
____ iconography	6) The construction and improvement of foundational services such as access to energy resources throughout the country.

Unit 4
Review Exercises

Check Your Understanding

Review the "Module Review" summaries at the end of each module in Unit 4 of your textbook. Make sure you understand all of the concepts before moving on to the next unit.

Unit 4 Multiple-Choice Question AP® Exam Practice

1. Which of the following best defines a sovereign state in the context of political geography?
   (A) A territory with complete cultural homogeneity
   (B) A political entity that possesses exclusive authority over its land and people
   (C) A region with shared linguistic characteristics
   (D) An area controlled by multiple political entities
   (E) A subnational unit within the United States such as Texas.

2. What factor most significantly contributed to the rise of nationalist movements among African civil servants during colonial rule?
   (A) Their exposure to Western education systems
   (B) Their desire for economic independence
   (C) Their opposition to European languages
   (D) Their access to military resources
   (E) Their control of local governments

3. Which of the following best explains how new communication technologies, such as the internet and social media, can challenge the power of the state?
   (A) By limiting public access to global information
   (B) By reinforcing state-controlled narratives in isolated regions
   (C) By enabling non-state actors to mobilize support and spread alternative viewpoints
   (D) By increasing government control over citizens' online behavior
   (E) By reducing cultural diversity through national media policies

4. According to the United Nations Convention on the Law of the Sea (UNCLOS), the Exclusive Economic Zone (EEZ) is defined as:
   (A) The area within 12 nautical miles of the shoreline
   (B) The area where international waters begin
   (C) The area within 24 nautical miles of the shoreline where coastal states have sovereignty
   (D) The area extending 200 nautical miles where coastal states have sole resource rights
   (E) The area designated for international shipping lanes

5. What is the primary purpose of reapportionment in the U.S. House of Representatives?
   (A) To redraw district boundaries within states
   (B) To ensure equal representation based on population changes
   (C) To manipulate voting outcomes for political advantage
   (D) To establish new voting precincts at the local level
   (E) To adjust state boundaries after each census

6. What is the primary characteristic of a unitary state?
   (A) Power is distributed among various subnational units
   (B) Power is concentrated in the central government
   (C) Power is shared equally between federal and state governments
   (D) Power is determined by autonomous regions
   (E) Power is regulated by international bodies

7. Which statement best describes how federalism affects Canada's relationship with Quebec?
   (A) It prevents any form of Quebec independence.
   (B) It eliminates French Canadian cultural identity.
   (C) It helps balance Quebec's independence demands with national unity.
   (D) It forces Quebec to abandon its unique status.
   (E) It grants Quebec complete sovereignty.

8. Which of the following best describes a failed state?
   (A) A state with strong economic development but weak cultural ties
   (B) A state whose political or economic system has become so weak that the government is no longer in control
   (C) A state with multiple ethnic groups living in harmony
   (D) A state with equal distribution of resources across regions
   (E) A state with strong military control but weak infrastructure

Unit 4 Free-Response Question AP® Exam Practice

Throughout history, the political organization of space has shifted due to changing notions of sovereignty, territoriality, and identity. In the contemporary world, political geography is shaped by globalization, internal divisions, and international cooperation. Use Figure 28.11 Partitioning of Africa after the Berlin Conference (p. 388) to answer the prompts below. Respond to parts A, B, C, D, E, F, and G.

A. Identify the type of boundaries established by the Berlin Conference in the map above.
B. Explain how the colonial boundaries shown on the map above have contributed to political instability or conflict in the contemporary world.
C. Define the concept of sovereignty in political geography
D. Describe the role of supranational organizations in reducing tensions between member states.
E. Explain how a choke point can increase the geopolitical importance of a region.
F. Explain the difference between a unitary state and a federal state.
G. Explain the degree to which physical geography can impact the effectiveness of a state's boundaries.

Notes:

Unit 5
Agricultural and Rural Land-Use Patterns and Processes

Unit Summary

In Unit 5, you will learn about how getting food has been the main way that humans have interacted with the physical environment for most of human history. We will dig into the complexities of agriculture, examine what geographers have learned about how humans get their food, and consider how humans can influence the sustainability of agriculture.

Modules in Unit 5

What do you see?

Look closely at the unit-opening photograph. Write down 5 things you observe about it. You will return to this image in the AP® Seeing Geography feature at the end of the unit.

1)

2)

3)

4)

5)

Module 33
Introduction to Agriculture

Before You Read the Module

Focus on Learning Goals

Use the Module learning goals as a way to divide or "chunk" the modules into parts. After you read a section of the module and answer the questions in this book, you should stop and ask yourself: "do I sufficiently understand the goal?"

LG 33-1: Explain how farming practices interact with the physical environment.
LG 33-2: Define climate and explain how it influences agriculture.
LG 33-3: Describe the different types of agricultural practices.

Key Terms

Take a look at the key terms you will encounter in this module.

agriculture
domesticated plant
domesticated animal
farmers
physical geography
nutrients
topography
climate
weather
tropical wet climate
tropical wet and dry climate
monsoon
monsoon rains
arid climate
semiarid (steppe) climate
moderate climate
humid subtropical climate
marine west coast climate
mediterranean climate
continental climate
humid continental climate
humid cold climate
intensive agriculture
subsistence agriculture
commercial agriculture
market gardening
truck farm
plantation
monocropping (monoculture)
mixed crop/livestock agriculture
cereal grains
millet
root crops
cash crop
paddy rice farming
upland or dry rice farming
grain farming
livestock fattening
feedlot
dairying
extensive agriculture
shifting cultivation
slash-and-burn (swidden) agriculture
nomadic herding (nomadic pastoralism or pastoralism)
livestock ranching

While You Read the Module

Answer the questions as you read Module 33: Introduction to Agriculture.

LG 33-1: *Explain how farming practices interact with the physical environment.*

1) Identify the three most important elements of physical geography for agriculture.

2) Describe characteristics of soils suitable for agriculture.

3) Explain how topography in Northern Ireland impacts the type of agriculture practiced there.

LG 33-2: *Define climate and explain how it influences agriculture.*

4) Explain how climate influences agriculture.

5) Describe how latitude influences distribution of climate types.

6) Complete the following table.

Climate Types	*Characteristics*
Tropical	
Dry	
Moderate	

Continental	

LG 33-3: ***Describe the different types of agricultural practices.***

7) Identify the two broad categories of agricultural practices.

8) Describe one difference between intensive subsistence agriculture and intensive commercial agriculture.

9) Complete the following table.

Intensive Agricultural Practice	*Description of Role the Physical Environment Plays*
Market gardening	
Plantation	
Mixed crop/livestock	
Rice	
Grain	

Livestock fattening	
Dairying	

10) Describe one major difference between intensive and extensive agriculture.

11) Complete the following table.

Extensive Agricultural Practice	*Description of Role the Physical Environment Plays*
Shifting cultivation	
Nomadic herding	
Livestock ranching	

After You Read the Module

Review the Key Terms

Match each key term on the left with its definition on the right.

____ agriculture

____ physical geography

1) The study of Earth's physical characteristics and processes: how they work, how they affect humans, and how humans affect them.

2) A climate with winter precipitation, unusually mild winters, and clear skies with abundant sunshine; found along the Mediterranean Sea and a few coastal regions.

____ climate	3) The average pattern of weather over a 30-year period for a particular region.
____ tropical wet climate	4) The planting and harvesting of domesticated plants and the raising of domesticated animals for food.
____ mediterranean climate	5) A climate located along the equator that experiences rain every day of the year.
____ intensive agriculture	6) The cultivation of a single commercial crop on extensive tracts of land.
____ subsistence agriculture	7) A fenced enclosure used for intensive livestock feeding that serves to limit livestock movement and associated weight loss.
____ commercial agriculture	8) Seeds that come from a wide variety of grasses cultivated around the world, including wheat, barley, sorghum, millet, oats, maize (corn), and rice.
____ market gardening	9) Food production mainly for consumption by the farming family and local community, rather than principally for sale in the market.
____ plantation	10) A highly mechanized commercial farming system that specializes in the production of cereal grains; requires large farms and widespread use of machinery, synthetic fertilizer, pesticides, and genetically engineered seeds.
____ monocropping (monoculture)	11) Farming oriented exclusively toward the production of agricultural commodities for sale in the market.
____ mixed crop/livestock agriculture	12) Crop cultivation and livestock rearing systems that use high levels of labor and capital relative to the size of the landholding.
____ cereal grains	13) A diversified system of agriculture based on the cultivation of cereal grains and root crops (such as potatoes and yams) and the rearing of herd livestock.

____ cash crop

____ grain farming

____ feedlot

____ dairying

____ extensive agriculture

____ shifting cultivation

____ nomadic herding (nomadic pastoralism or pastoralism)

____ livestock ranching

14) A small-scale farming system in which a farmer plants one to a few acres that produce a diverse mixture of vegetable and fruits, mostly for sale in local and regional markets.

15) A crop raised to be sold for profit rather than to feed the farm family and the livestock; common cash crops are cotton, flax, hemp, coffee, and tobacco.

16) Large landholding devoted to capital-intensive, specialized production of a single tropical or subtropical crop for the global marketplace.

17) A system of breeding and rearing herd livestock, such as cattle, sheep, or goats, by following the seasonal movement of rainfall to areas of open pasturelands.

18) The practice of using extensive tracts of land to rear herds of livestock to sell as meat, hides, or wool.

19) Crop cultivation and livestock rearing systems that require little hired labor or monetary investment to successfully raise crops and animals.

20) A farming system that specializes in the breeding, rearing, and utilization of livestock (primarily cows) to produce milk and its various by-products, such as yogurt, butter, and cheese.

21) The cultivation of a plot of land until it becomes less productive, typically over a period of about three to five years; when productivity drops, the farmer shifts to a new plot of land that has been prepared by slash-and-burn agriculture.

Module 34
Settlement Patterns and Survey Methods

Before You Read the Module

Focus on Learning Goals

Use the Module learning goals as a way to divide or "chunk" the modules into parts. After you read a section of the module and answer the questions in this book, you should stop and ask yourself: "do I sufficiently understand the goal?"

LG 34-1: Explain the effects of major agricultural practices on the rural landscape.
LG 34-2: Identify the three primary rural settlement patterns.
LG 34-3: Explain the three rural survey methods.

Key Terms

Take a look at the key terms you will encounter in this module.

rural area
rural settlement
agricultural landscape
silo
intercropping
tundra
settlement patterns
clustered settlement (farm village)
farmstead
dispersed settlement (isolated settlement pattern)
linear settlement pattern
survey methods
cadastral survey
metes and bounds
township and range
long-lot survey method

While You Read the Module

Answer the questions as you read Module 34: Settlement Patterns and Survey Methods.

LG 34-1: ***Explain the effects of major agricultural practices on the rural landscape.***

1) Identify the percentage of Earth used for the production of food.

2) Identify three distinctive features of agricultural landscapes.

3) Compare the agricultural landscape of some areas of central New York state to the Northern Plains region of the United States and Canada.

4) Identify three factors that influence the agricultural landscape for market gardening and truck farming.

5) Complete the following table.

Intensive Agricultural Practice	*Impact on Agricultural Landscape*
Market Gardening and Truck Farming	
Plantations	
Mixed Crop/Livestock Farming	
Paddy Rice Farming	
Upland Rice Farming	
Grain Farming	
Livestock Fattening	
Dairy Farming	

6) Complete the following table.

Extensive Agricultural Practice	*Impact on Agricultural Landscape*
Shifting cultivation	
Nomadic herding	
Livestock ranching	

LG 33-2: ***Identify the three primary rural settlement patterns.***

7) Explain a historical reason clustered settlements or farm villages are the most common type of settlement pattern.

8) Describe conditions that foster dispersed or isolated rural settlements.

9) Describe a political reason that influenced the development of linear settlement patterns.

LG 33-3: ***Explain the three rural survey methods.***

10) Describe how physical features delineate property boundaries in the metes and bounds survey system.

11) Explain how the township and range system influenced settlement patterns in much of the western half of the United States.

12) Explain an advantage of the long-lot survey method for farmers.

After You Read the Module

Review the Key Terms

Match each key term on the left with its definition on the right.

____ rural area	1) A settlement pattern in which buildings are arranged in a line, often along a road or river; limited to areas where legal systems dictated that property lines must be rectangular.
____ rural settlement	2) A unit-block surveying method whose basic unit is a rectangle that is typically 10 times longer than it is wide.
____ settlement patterns	3) A tightly bunched farm settlement that has anywhere from a few dozen to several hundred inhabitants.
____ clustered settlement or farm village	4) Area located outside of towns and cities; all the space, population, and housing not included in an urban area.
____ dispersed settlement or isolated settlement pattern	5) A small group of people living outside of an urban area.
____ linear settlement pattern	6) Land survey method created by the U.S. Land Ordinance of 1785, which divides most of the country's territory into a grid of square-shaped townships with 6-mile sides.
____ survey methods	7) A settlement pattern in which families live relatively distant from one another.
____ metes and bounds	8) The methods used by surveyors to lay out property lines.
____ township and range	9) The ways in which people organize themselves on the land.
____ long-lot survey method	10) Survey method that uses natural features such as trees, boulders, and streams to delineate property boundaries.

Module 35
Agricultural Origins and Diffusions

Before You Read the Module

Focus on Learning Goals

Use the Module learning goals as a way to divide or "chunk" the modules into parts. After you read a section of the module and answer the questions in this book, you should stop and ask yourself: "do I sufficiently understand the goal?"

LG 35-1: Define domestication.
LG 35-2: Identify the early hearths of domesticated plants and animals.
LG 35-3: Explain the global patterns of diffusion for domesticated plants and animals.

Key Terms

Take a look at the key terms you will encounter in this module.

domestication
First Agricultural Revolution
teosinte
Mesoamerica
biodiversity
hearth
Fertile Crescent
Indus River Valley
Columbian Exchange

While You Read the Module

Answer the questions as you read Module 35: Agricultural Origins and Diffusions.

LG 35-1: ***Define domestication.***

1) Define *domestication.*

2) Explain why women probably played a larger role than men in early plant domestication.

3) Explain why Carl Sauer believed that the original areas of plant domestication were in hilly regions.

LG 35-2: ***Identify the early hearths of domesticated plants and animals.***

4) Identify how long ago the earliest domestication of plants probably began according to archaeological research.

5) Complete the following table.

Agricultural Hearth	*Key Crops and Animals*
Southwest Asia	
China, South and Southeast Asia, and Malaysia	
Africa	
The Americas	

LG 35-3: *Explain the global patterns of diffusion for domesticated plants and animals.*

6) Using Figure 35.7 (p. 508), describe how the Columbian Exchange resulted in the global spread of various domesticated plants and animals.

7) Explain how the diffusion of African rice demonstrates the importance of Indigenous knowledge.

8) Explain how acceptance of a new crop can have unintended consequences.

After You Read the Module

Review the Key Terms

Match each key term on the left with its definition on the right.

____ domestication	1) Area in Southwest Asia that includes the river valleys of the Tigris and Euphrates; the earliest center for domestication of seed plants.
____ First Agricultural Revolution	2) The cultural region in the Americas that includes the diverse civilizations in the modern-day countries of Mexico, Guatemala, Honduras, Belize, El Salvador, Nicaragua, and Costa Rica.
____ teosinte	3) The interaction and widespread transfer of plants, animals, culture, human populations, technology, disease, and ideas between the Americas, Europe, Africa, and Asia in the fifteenth and sixteenth centuries.
____ Mesoamerica	4) A center where innovations or new practices develop and from which the innovations or new practices spread or diffuse.
____ biodiversity	5) The period during which the early domestication and diffusion of plants and animals and the cultivation of seed crops led to the development of agriculture.
____ hearth	6) The long-term process through which humans selectively breed, protect, and care for individuals taken from populations of wild plant and animal species to create genetically distinct species, known as domesticates.
____ Fertile Crescent	7) Large wild grass native to Mexico that produced the small ears of maize (corn) that were a favored food among early groups in Mesoamerica.
____ Indus River Valley	8) The variety and variability among species and ecosystems.
____ Columbian Exchange	9) Area along the Indus River that flows from the highlands of Tibet and continues down along the border between present-day Pakistan and India; a site of the earliest domestication of plants and herd animals.

Module 36

The Second Agricultural Revolution

Before You Read the Module

Focus on Learning Goals

Use the Module learning goals as a way to divide or "chunk" the modules into parts. After you read a section of the module and answer the questions in this book, you should stop and ask yourself: "do I sufficiently understand the goal?"

LG 36-1: Explain how technologies and innovations increased agricultural productivity during the Second Agricultural Revolution.

LG 36-2: Describe how increases in agricultural productivity affected societies during the Second Agricultural Revolution.

Key Terms

Take a look at the key terms you will encounter in this module.

Second Agricultural Revolution	agrichemicals
seed drill	synthetic fertilizer
mechanical reaper	pesticide
scythe	herbicide

While You Read the Module

Answer the questions as you read Module 36: The Second Agricultural Revolution.

LG 36-1: ***Explain how technologies and innovations increased agricultural productivity during the Second Agricultural Revolution.***

1) Explain how the Columbian Exchange led to the cultivation of previously unused land.

2) Explain how the seed drill increased agricultural efficiency and decreased waste.

3) Explain why use of the mechanical reaper rapidly diffused around the world.

4) Explain how railroads and canals provided farmers with affordable access to distant markets.

5) Explain how agrichemicals impacted agricultural production during the Second Agricultural Revolution.

LG 36-2: ***Describe how increases in agricultural productivity affected societies during the Second Agricultural Revolution.***

6) Describe an impact of better diets on human populations during the Second Agricultural Revolution.

7) Describe how the shift from primarily subsistence farming to the production of cash crops impacted rural social structure during the Second Agricultural Revolution.

After You Read the Module

Review the Key Terms

Match each key term on the left with its definition on the right.

____ Second Agricultural Revolution	1) A machine used to harvest grain crops mechanically; patented by Cyrus McCormick in 1831.
____ seed drill	2) Period that brought improved methods of cultivation, harvesting, and storage of farm produce that began in the late 1600s and continued through the 1930s.
____ mechanical reaper	3) A machine for planting seeds in a row.
____ scythe	4) Chemical substance designed to kill or inhibit the growth of unwanted plants (weeds) that compete with crops.
____ agrichemicals	5) Chemical compounds obtained from petroleum and natural gas for use in agriculture; agrichemicals include fertilizers, pesticides, and herbicides.
____ synthetic fertilizer	6) Material used to kill or repel animals or insects that can damage, destroy, or inhibit crop growth.
____ pesticide	7) An agricultural hand tool with a curved blade used for cutting grain in the fields.
____ herbicide	8) Industrially manufactured nitrogen, phosphorus, and potassium, made from petroleum by-products; contains higher concentrations of nutrients for plants than natural fertilizers.

Module 37
The Green Revolution

Before You Read the Module

Focus on Learning Goals

Use the Module learning goals as a way to divide or "chunk" the modules into parts. After you read a section of the module and answer the questions in this book, you should stop and ask yourself: "do I sufficiently understand the goal?"

> **LG 37-1:** Describe the three primary characteristics of the Green Revolution.
>
> **LG 37-2:** Explain the positive consequences of the Green Revolution on human populations and the environment.
>
> **LG 1-3:** Explain the negative consequences of the Green Revolution on human populations and the environment.

Key Terms

Take a look at the key terms you will encounter in this module.

Green Revolution	sorghum
crossbreeding	endemic
hybrid	environmental contamination
double-cropping	soil salinization
multicropping	soil salinity
cassava	orphan crop

While You Read the Module

Answer the questions as you read Module 37: The Green Revolution.

LG 37-1: ***Describe the three primary characteristics of the Green Revolution.***

1. Describe how advanced methods of plant breeding developed in the mid-twentieth century impacted crop yields.

2. Identify the three crops whose yields were impacted first by Green Revolution techniques.

3. Explain why some farmers were not able to benefit from the Green Revolution.

LG 37-2: ***Explain the positive consequences of the Green Revolution on human populations and the environment.***

4. Describe how increased crop yields during the Green Revolution impacted the economic development at the national scale of some countries such as Thailand and India.

5. Compare World Bank estimates of world hunger in 1950 to those today.

6. Describe how the Green Revolution impacted the expansion of land used for agriculture on a global scale.

LG 37-3: ***Explain the negative consequences of the Green Revolution on human populations and the environment.***

7. Explain how the Green Revolution in India is an example of hierarchical diffusion.

8. Explain how expansion of the production of cash crops can impact production of food crops on a local scale.

9. Explain why the Green Revolution had less of an impact on Africa than on other geographic regions.

10. Describe negative environmental impacts of irrigation associated with Green Revolution techniques.

Soil salinity	
Groundwater depletion	

11. Explain why scientists are concerned about the impact of global warming in the context of a growing world population.

12. Explain how increasing production of orphan crops may impact food and economic security on a regional scale.

After You Read the Module

Review the Key Terms

Match each key term on the left with its definition on the right.

_____ Green Revolution	1) A root vegetable native to South America.
_____ crossbreeding	2) Planting two or three crops per year on the same land.
_____ hybrid	3) The U.S.-supported development of high-yield seed varieties that increased the productivity of cereal crops and accompanying agricultural technologies for transfer to less developed countries.
_____ double-cropping	4) A grain plant native to northeast Africa.
_____ multicropping	5) The process of mixing different species or varieties of plants or animals to produce hybrids.
_____ cassava	6) The offspring of two plants or animals of different species or varieties.
_____ sorghum	7) Planting another crop on the same plot of land as soon as the first crop has been harvested.
_____ endemic	8) The concentration of dissolved salts in the soil.

____ environmental contamination

9) Native to or characteristic of a certain environment.

____ soil salinization

10) Crop not typically traded on global markets but that plays an important role in regional food security.

____ soil salinity

11) Chemical residue that builds up with each application of synthetic fertilizers and pesticides.

____ orphan crop

12) A measure of the concentration of dissolved salts in the soil; high soil salinity results from poor irrigation practices.

Module 38
Agricultural Practices and Economic Forces

Before You Read the Module

Focus on Learning Goals

Use the Module learning goals as a way to divide or "chunk" the modules into parts. After you read a section of the module and answer the questions in this book, you should stop and ask yourself: "do I sufficiently understand the goal?"

LG 38-1: Explain how subsistence and commercial agricultural practices are distributed across the world.

LG 38-2: Describe how bid-rent theory explains the relationship between land cost and intensive and extensive agricultural practices.

LG 38-3: Explain how large-scale commercial agriculture affects family farms.

LG 38-4: Explain how commodity chains affect agricultural practices.

LG 38-5: Explain how technology has increased economies of scale and carrying capacity in agriculture.

Key Terms

Take a look at the key terms you will encounter in this module.

capital expenditures
bid-rent theory
central business district (CBD)
large-scale commercial operation
agricultural cooperative
family farm
commodity
commodity chain
agribusiness
agricultural carrying capacity
concentrated animal feeding operation (CAFO)
cool chain

While You Read the Module

Answer the questions as you read Module 38: Agricultural Practices and Economic Forces.

LG 38-1: *Explain how subsistence and commercial agricultural practices are distributed across the world.*

1) Identify the level of economic development that is predominant in the regions in which subsistence agricultural production is found.

2) Identify the climate type associated with fig, date, olive, and grape farming in Figure 38.2 (p. 538).

LG 38-2: ***Describe how bid-rent theory explains the relationship between land cost and intensive and extensive agricultural practices.***

3) Explain how high competition for land close to the market impacts the price per unit of land.

4) Explain why farmers who engage in intensive farming practices such as dairy farming and market gardening are willing to purchase land at a higher cost per acre than farmers who engage in extensive agricultural practices such as potato farming.

LG 38-3: ***Explain how large-scale commercial agriculture affects family farms.***

5) Explain why the number of dairy farms in the United States has decreased since 1935.

LG 38-4: ***Explain how commodity chains affect agricultural practices.***

6) Explain how each link in a commodity chain impacts the value of the commodity.

7) Using Figure 38.10 (p. 543), identify the five main sectors of the commodity chain.

8) Explain why most of the chicken supply in the United States is handled by a half dozen very large corporations such as Tyson and Perdue.

LG 38-5: ***Explain how technology has increased economies of scale and carrying capacity in agriculture.***

9) Explain how the concept of economies of scale influences smaller farms to join together in farm cooperatives.

10) Identify the percent of Earth's land that is considered unsuitable for agriculture.

11) Describe industrial inputs that have increased agricultural carrying capacity by allowing more animals to be raised in a smaller area.

12) Explain how cool chains have impacted the spatial relationship between where food is produced and where it is sold.

After You Read the Module

Review the Key Terms

Match each key term on the left with its definition on the right.

____ capital expenditures	1) An organization where farmers pool their resources in certain areas of activity such as services or production; services or production resources are provided to individual farm members.
____ bid-rent theory	2) Assets that cost money, such as land, machinery, synthetic fertilizers and pesticides, seeds, and livestock feed.
____ central business district (CBD)	3) A dense cluster of offices and shops located at a city's most accessible point, usually its center.

____ large-scale commercial operations	4) A farming operation majority-owned by a family or family corporation that sells its products to some defined market, either directly or through a cooperative.
____ agricultural cooperative	5) Explains how the demand for and price of land decrease as its distance from the central business district increases.
____ family farm	6) A large-scale farm oriented exclusively toward the production of agricultural commodities for sale in the market.
____ commodity	7) A series of links connecting a commodity's many places of production, distribution, and consumption.
____ commodity chain	8) In agriculture, a primary product that can be bought and sold, such as coffee, rice, or milk.
____ agribusiness	9) Animal rearing system that confines livestock (such as cattle, sheep, turkeys, chickens, and hogs) in high-density cages only large enough to allow the animal's body to grow and to accommodate equipment for feeding and waste removal.
____ agricultural carrying capacity	10) The system that uses refrigeration and food-freezing technologies to keep farm produce fresh in climate-controlled environments at every stage of transport from field to retail grocers and restaurants.
____ concentrated animal feeding operation (CAFO)	11) Refers to the amount of land that is available in a given region that is available for farming.
____ cool chain	12) Large corporation that provides a vast array of goods and services to support the agricultural industry.

Module 39
The von Thünen Model

Before You Read the Module

Focus on Learning Goals

Use the Module learning goals as a way to divide or "chunk" the modules into parts. After you read a section of the module and answer the questions in this book, you should stop and ask yourself: "do I sufficiently understand the goal?"

> **LG 39-1:** Identify the underlying assumptions of the von Thünen model and explain how it applies to historic patterns of agricultural land use.
>
> **LG 39-2:** Describe how the von Thünen model explains contemporary patterns of agricultural production at different scales.

Key Terms

Take a look at the key terms you will encounter in this module.
hinterland

While You Read the Module

Answer the questions as you read Module 39: The von Thünen Model.

LG 39-1: ***Identify the underlying assumptions of the von Thünen model and explain how it applies to historic patterns of agricultural land use.***

1) Identify the factor von Thünen determined was key to understanding the distribution and intensity of agriculture.

2) Describe the six assumptions underlying von Thünen's model.

3) Explain why it is important to understand that von Thünen developed his model in the 1820s based on agriculture in northwest Germany.

4) Using Figure 39.1 (p. 552), sketch and label von Thünen's model.

5) Explain why transportation costs to the market for livestock that would be slaughtered were minimal in von Thünen's model.

6) Explain how the presence of a navigable river could change the location of dairy farming.

7) Explain how agricultural products raised by farmers in the hinterlands of New York City in the 1820s reflect patterns predicted by von Thünen's model.

LG 39-2: ***Describe how the von Thünen model explains contemporary patterns of agricultural production at different scales.***

8) Using Figure 39.5 (p. 555), explain how climate plays a role in the placement of specialty crops such as grapes, olives, and citrus fruits when applying von Thünen's model on a national scale to the United States.

9) Explain how looking at maps of agricultural production on the regional and local scales can reveal variations in data different from what can be seen in maps at the national scale.

10) Use Figure 39.9 (p. 558) to explain how climate might impact agricultural production regions in Europe.

After You Read the Module

Review the Key Terms

TRUE or FALSE: A hinterland is the area surrounding a city.

Module 40

The Global System of Agriculture

Before You Read the Module

Focus on Learning Goals

Use the Module learning goals as a way to divide or "chunk" the modules into parts. After you read a section of the module and answer the questions in this book, you should stop and ask yourself: "do I sufficiently understand the goal?"

> **LG 40-1:** Explain the interdependence of the global supply chain and agricultural production and consumption.
>
> **LG 40-2:** Explain export commodities and how they affect countries.
>
> **LG 40-3:** Explain the effects of political relationships, infrastructure, and patterns of world trade on global food distribution networks.

Key Terms

Take a look at the key terms you will encounter in this module.

global supply chain	export commodity
contract farming	subsidies
proprietary seeds	famine

While You Read the Module

Answer the questions as you read Module 40: The Global System of Agriculture.

LG 40-1: ***Explain the interdependence of the global supply chain and agricultural production and consumption.***

1) Identify the process that increased the interdependence of the production, consumption, and globalization of food.

2) Describe the role a farmer plays in the global supply chain for chicken.

3) Explain why the U.S. share of the global poultry market is shrinking in the twenty-first century.

LG 40-2: ***Explain export commodities and how they affect countries.***

4) Explain how the production of export commodities impacts crops produced for consumption by people in the local area.

5) Explain how growth in the production of sugarcane in Brazil and India impacted countries in the Caribbean, many of which were dependent on exporting sugarcane to western Europe in the twentieth century.

6) Describe environmental impacts of the production of sugarcane on islands in the Caribbean.

7) Explain how comparatively wealthy consumers in more industrialized countries contribute to food insecurity of people in less industrialized countries.

LG 40-3: ***Explain the effects of political relationships, infrastructure, and patterns of world trade on global food distribution networks.***

8) Explain why international political economics is the primary cause of hunger in the world.

9) Explain how internal government policies in China in the mid-twentieth century led to famine.

10) Explain how infrastructure built by colonial powers has impacted food distribution to people within particular countries or regions such as India, Mexico, Africa, and Southeast Asia.

11) Explain why international trade is likely to play an increasingly important role in meeting global food demand in the coming decades.

12) Describe how global trade in coffee has fluctuated since the 1960s.

After You Read the Module

Review the Key Terms

Match each key term on the left with its definition on the right.

____ global supply chain	1) A cash crop that is produced for export to wealthier countries at the expense of crop production for local consumption.
____ contract farming	2) Seeds that are developed and entirely owned by a company.
____ proprietary seeds	3) Arrangement between an independent farmer and an agribusiness company to produce a crop; the agribusiness provides the farmer with all the supplies needed to produce a crop in exchange for a guaranteed price and buyer.
____ export commodity	4) Extreme scarcity of food.
____ subsidies	5) Agribusinesses, organized at the global scale; encompasses all elements of growing, harvesting, processing, transporting, marketing, consuming, and disposing of food for people.
____ famine	6) Guaranteed prices for staple food crops.

Module 41

Consequences of Agricultural Practices

Before You Read the Module

Focus on Learning Goals

Use the Module learning goals as a way to divide or "chunk" the modules into parts. After you read a section of the module and answer the questions in this book, you should stop and ask yourself: "do I sufficiently understand the goal?"

LG 41-1: Explain how agricultural practices and land use impact the environment.
LG 41-2: Explain how agricultural practices alter the landscape.
LG 41-3: Explain how changes in agricultural practices affect society.

Key Terms

Take a look at the key terms you will encounter in this module.

runoff
nutrient pollution
dead zones
deforestation
desertification
water control land reclamation
irrigated agriculture
water mining
aquifer
biofuel
biodiesel

While You Read the Module

Answer the questions as you read Module 41: Consequences of Agricultural Practices.

LG 41-1: ***Explain how agricultural practices and land use impact the environment.***

1) Complete the following table.

Agricultural Practice	*Effects of Agricultural Practice*
Use of manufactured pesticides	
Irrigation	

2) Explain the loss of forest cover in central Europe between 900 and 1900 CE as seen in Figure 41.4 (p. 577).

3) Explain how deforestation contributes to climate change.

4) Describe the types of physical regions most at risk of desertification.

5) Describe impacts of conservation efforts.

LG 41-2: ***Explain how agricultural practices alter the landscape.***

6) Describe how shifting cultivation can alter the landscape.

7) Describe how the Hani people of China altered the physical environment to create paddy rice terraces.

8) Describe how competition from sedentary agriculture has impacted the ability of nomadic pastoralists to minimize damage to the physical environment.

9) Explain how draining wetlands alters the physical environment.

10) Identify two technologies that facilitate the capture and delivery of water for use in irrigated agriculture.

LG 41-3: ***Explain how changes in agricultural practices affect society.***

11) Describe changes in western food preferences since the 1960s.

12) Describe the impact of the mechanization of agriculture on women.

13) Describe the purpose of the 2007 Energy Independence and Security Act.

14) Describe a potential environmental impact of increased biofuel use.

15) Describe two potential drawbacks of increased biofuel use.

After You Read the Module

Review the Key Terms

Match each key term on the left with its definition on the right.

____ runoff	1) The process by which once-fertile land becomes desert as a result of climate variation or human activities
____ nutrient pollution	2) Clearing and destruction of forests to clear land for agriculture use
____ dead zones	3) Consequence of overuse of fertilizer; occurs when excess nutrients seep down into groundwater or are carried into nearby waterways as runoff
____ deforestation	4) The flow of rain or irrigation water over land
____ desertification	5) The process of draining land inundated with either fresh water or salt water to increase areas for agricultural production
____ water control land reclamation	6) Sections of a body of water where there is very little aquatic life
____ irrigated agriculture	7) Underground water deposited hundreds of thousands of years ago
____ water mining	8) Fuel made from vegetable oils
____ aquifer	9) The use of deep-well drilling technology and powerful industrial pumping systems to remove water in the ground
____ biofuel	10) A fuel derived from organic wastes or plant materials
____ biodiesel	11) Farming that relies on the controlled application of water to cultivated fields

Module 42
Challenges of Contemporary Agriculture

Before You Read the Module

Focus on Learning Goals

Use the Module learning goals as a way to divide or "chunk" the modules into parts. After you read a section of the module and answer the questions in this book, you should stop and ask yourself: "do I sufficiently understand the goal?"

LG 42-1: Explain whether contemporary agricultural innovations are sustainable.
LG 42-2: Describe how food choice movements influence patterns of food production and consumption.
LG 42-3: Explain the contemporary challenges of feeding the global population.
LG 42-4: Explain why large-scale agribusinesses have an advantage over small-scale farmers.

Key Terms

Take a look at the key terms you will encounter in this module.

sustainable agriculture
genetically modified organism (GMO)
aquaculture
mariculture
urban farming
community-supported agriculture
farmers' market
organic farming
conventional agriculture
value-added specialty crop
fair trade
slow-food
locavores
food desert
food security
food insecurity

While You Read the Module

Answer the questions as you read Module 42: Challenges of Contemporary Agriculture.

LG 42-1: ***Explain whether contemporary agricultural innovations are sustainable.***

1) Identify two contemporary agricultural innovations with challenges related to long-term sustainability.

2) Identify the year in which commercial production of foods made with genetically modified organisms (GMOs) began in the United States.

3) Describe two benefits of GMOs for farmers.

4) Although most scientists agree that genetically modified food is safe, explain why there is debate about their environmental sustainability.

5) Describe an environmental impact of commercial aquaculture in tropical regions.

LG 42-2: ***Describe how food choice movements influence patterns of food production and consumption.***

6) Complete the following table.

Food Choice Movement	*Influence on Patterns of Food Production and Consumption*
Urban farming	
Community-supported agriculture	
Organic farming	
Value-added specialty crops	
Fair trade	
Local food movement	

LG 42-3: ***Explain the contemporary challenges of feeding the global population.***

7) Identify the type of urban neighborhoods most likely to be considered food deserts.

8) Explain how political conflict contributes to food insecurity.

9) Explain why food safety affects both poor and wealthy populations.

10) Explain how global climate change presents a challenge to feeding the global population.

11) Identify the process that causes farmland to be lost to housing and retail developments in metropolitan areas.

LG 42-4: ***Explain why large-scale agribusinesses have an advantage over small-scale farmers.***

12) Describe the population who currently benefits the most from the contemporary global network of food production.

13) Explain how agribusiness has an advantage over small-scale farmers in the context of government policies and food production.

After You Read the Module

Review the Key Terms

Match each key term on the left with its definition on the right.

____ sustainable agriculture	1) The cultivation and harvesting of aquatic organisms under controlled conditions
____ genetically modified organism (GMO)	2) A commitment to satisfying human food and textile needs and to enhancing the quality of life for farmers and society as whole, now and in the future; it requires a balance among feeding the growing population, minimizing environmental impacts, and ensuring social justice
____ aquaculture	3) A living organism, including crops and livestock, that is produced through genetic engineering
____ mariculture	4) The farming of saltwater species such as shrimp, oysters, and marine fish
____ urban farming	5) The practice of growing fruits and vegetables on small private plots or shared community gardens within the confines of a city
____ community-supported agriculture (CSA)	6) A direct-to-consumer marketing arrangement in which farmers are guaranteed buyers for their produce at guaranteed prices and consumers receive fresh food directly from the producers
____ farmers' market	7) The production of crops and livestock using ecological processes, natural biodiversity, and renewable resources rather than industrial practices and synthetic inputs
____ organic farming	8) A venue (ranging from a few stalls in the street to covered enclosures extending a few city blocks) in which farmers sell their produce directly to consumers
____ conventional agriculture	9) Farming that depends on manufactured synthetic inputs, GMO seeds, and other industrial practices

____ value-added specialty crop	10) Movement that resists fast food by preserving the cultural cuisine and the associated food and farming practices of an ecoregion
____ fair trade	11) A crop whose physical state or form has been changed
____ slow-food	12) A certification program that supports good crop prices for farmers and environmentally sound farming practices
____ locavores	13) Area with limited access to fresh, nutritious foods
____ food desert	14) People who dedicate themselves to slow-food diets and to obtaining as much of their nutrition as possible from local farmers
____ food security	15) Occurs when large numbers of people experience long periods of inadequate diets
____ food insecurity	16) According to the United Nations, the situation in which all people, at all times, have physical and economic access to enough safe and nutritious food that meets their dietary needs and food preferences for an active and healthy life

Module 43
Women in Agriculture

Before You Read the Module

Focus on Learning Goals

Use the Module learning goals as a way to divide or "chunk" the modules into parts. After you read a section of the module and answer the questions in this book, you should stop and ask yourself: "do I sufficiently understand the goal?"

LG 43-1: Explain how the percentage of women in the agricultural labor force varies worldwide.
LG 43-2: Explain how women's roles differ between subsistence and commercial agriculture.

Key Terms

Take a look at the key terms you will encounter in this module.
division of labor

While You Read the Module

Answer the questions as you read Module 43: Women in Agriculture.

LG 43-1: ***Explain how the percentage of women in the agricultural labor force varies worldwide.***

1) Describe two limitations of the data related to women's roles in agriculture when attempting to make comparisons across regions.

2) Explain how the type of subsistence farming that is practiced can influence women's farm labor.

3) Describe the difference in time spent in agricultural production between women in developed countries and women in developed countries.

LG 43-2: ***Explain how women's roles differ between subsistence and commercial agriculture.***

4) Identify the two factors that typically influence the division of labor in societies in which many people are engaged in subsistence agriculture.

5) Describe the role women generally play in mixed crop/livestock farming in developing countries.

6) Explain how globalization has influenced the role of some women in the tropical regions in which shifting cultivation is practiced.

7) Describe two differences in the division of labor between men and women in regions in which paddy rice farming is common.

8) Explain how the relatively weak social and economic position of women in traditional pastoralist societies impacts their role in agriculture.

9) Describe one similarity between farms owned by women in the European Union and those owned by women in the United States.

10) Explain how empowering women is likely to impact food security on a local level.

After You Read the Module

Review the Key Terms

TRUE or FALSE: In subsistence systems, division of labor is traditionally based only on age.

Unit 5
Review Exercises

Check Your Understanding

Review the "Module Review" summaries at the end of each module in Unit 5 of your textbook. Make sure you understand all of the concepts before moving on to the next unit.

Unit 5 Multiple-Choice Question AP® Exam Practice

1. Identify which type of agriculture is evident in Figure 33.14 (p. 483).
   (A) Grain farming
   (B) Dairying
   (C) Pastoralism
   (D) Shifting cultivation
   (E) Intensive

2. Which best explains why clustered settlement patterns are found throughout the world?
   (A) Farm villages enabled farmers to defend themselves during times of insecurity.
   (B) Farm villages provided the greatest degree of privacy.
   (C) Clustered settlements facilitated the farming of well-drained land where water is readily available.
   (D) Clustered settlements were favored by local governments who mandated that people live in farm villages.
   (E) Farm villages promoted the physical health of farmers, who walked out of the village each day to work the land.

3. Which best explains the purpose of the township and range survey method in the United States?
   (A) To facilitate farmers' access to waterways
   (B) To avoid property disputes that were common in the metes and bounds system
   (C) To impose a checkerboard pattern on the agricultural landscape of the western United States
   (D) To provide farmers with access to the national railroad network and highway system
   (E) To promote equitable land ownership by small-scale independent farmers

4. Identify the agricultural hearth of maize (corn).
   (A) Indus River Valley
   (B) West Africa
   (C) Southeast Asia
   (D) Mesoamerica
   (E) Fertile Crescent

5. Which of the following best describes how the Erie Canal impacted farm productivity after it opened in 1825?
   (A) It had virtually no impact because most farmers could not afford the rates required to access the canal.

(B) It increased farm productivity by reducing the amount of water farmers needed to cultivate their crops.
(C) It decreased farm productivity by increasing the movement of people from farm to market.
(D) It increased farm productivity by helping farmers spend less time engaged in manual labor on their farms.
(E) It increased farm productivity by allowing farmers to transport nonperishable products much faster and at a much lower cost than was previously possible.

6. Which best describes a limitation of the Green Revolution?
(A) The increase in productivity allowed for the development of an export economy in Asian and Latin American countries.
(B) The Green Revolution has led to increased crop productivity for cereal grains grown by farmers using high-yield seeds.
(C) The geographical impact of the Green Revolution has been highly varied with some regions benefiting much less than others.
(D) The Green Revolution substantially increased world hunger in some world regions.
(E) By increasing production on the same acreage, the Green Revolution made the use of agricultural land and resources more efficient.

7. Which best explains how cool chains have impacted commodity chains?
(A) The price of all commodities has gone down because of advances in communication.
(B) Distributors must be located closer to processors because advances in climate-controlled transportation have increased the price of distributing products.
(C) The number of sectors in commodity chains has decreased due to advances in climate-controlled transportation.
(D) Many products can now be consumed much farther from where they were produced because of advances in climate-controlled transportation.
(E) Each sector in a commodity chain adds roughly the same value to a commodity as compared to before cool chains when producers added much more value to a commodity than processors, distributors, or retailers.

8. Which of the following is a major limitation of the von Thünen model of agricultural production as it applies to contemporary agriculture?
(A) Change over time in the model to incorporate advances in technology such as improvements in water transportation
(B) Its incorporation of agricultural activity in the hinterland
(C) Its depiction of the spatial distribution of types of commercial agriculture
(D) Its assumption that there is only one market
(E) Its inability to identify regions of specialty agricultural production

9. By the early twentieth century, India had a rail network over 25,000 miles in length. This system was built by colonial powers primarily to move export commodities to ports. The construction of this rail system best illustrates which factor related to the global system of agriculture?
(A) Dependency on export commodities

(B) Patterns of world trade
(C) Infrastructure
(D) Political relationships
(E) The impact of agricultural subsidies

10. Located on the arid borderland between Kazakhstan and Uzbekistan in central Asia, the Aral Sea is just a fraction of its former size as one of the four largest lakes in the world in the mid-twentieth century. The change over time in the size of the Aral Sea illustrates the impact on the landscape of which agricultural practice?
(A) Terraced farming
(B) Irrigation
(C) Draining wetlands
(D) Shifting cultivation
(E) Deforestation

11. Which most accurately describes a drawback of organic and fair trade products for consumers in the United States?
(A) They generally cost more for consumers because neither is produced in the United States, and the cost of importing them results in higher prices for consumers.
(B) Both require the use of genetically modified organisms, and this results in a higher price for consumers.
(C) Both are produced with high amounts of synthetic chemical fertilizers and pesticides, which can impact the health of consumers.
(D) Neither can be legally produced in the United States, and the cost of importing them results in higher prices for consumers.
(E) Both are typically more expensive to produce, and the cost is usually passed on to the consumer.

Unit 5 Free-Response Question AP® Exam Practice

1. Agricultural production and consumption patterns vary in different locations, presenting different environmental, social, economic, and cultural opportunities and challenges. Respond to parts A, B, C, D, E, F, and G.
    A. Define land cover change.
    B. Describe a global impact of deforestation in tropical regions.
    C. Describe a possible impact of conservation efforts.
    D. Explain ONE environmental challenge associated with contemporary commercial aquaculture.
    E. Explain how the production of cash crops for export can impact food production for the local population.
    F. Explain why food choice movements vary according to local conditions.
    G. Explain the degree to which food insecurity is more likely to impact people living in regions less impacted by techniques associated with the Green Revolution.

Notes:

Unit 6
Cities and Urban Land-Use Patterns and Processes

Unit Summary

Imagine walking through global cities like New York or Bangkok. You might encounter a mix of towering skyscrapers and traditional architecture. However, beyond the impressive facades, these urban centers can also harbor issues like traffic, pollution, and socioeconomic disparities. The increasing movement of people to cities, known as urbanization, is creating numerous megacities and metacities with over ten million inhabitants. These large urban areas are central to globalization, driving interconnectedness and interdependence worldwide.

Modules in Unit 6

Module 44: The Origin and Influences of Urbanization
Module 45: Cities Across the World
Module 46: Cities and Globalization
Module 47: The Size and Distribution of Cities
Module 48: The Internal Structure of Cities
Module 49: Density, Land Use, and Infrastructure
Module 50: Urban Sustainability
Module 51: Challenges of Urban Changes
Module 52: Challenges of Urban Sustainability

What do you see?

Look closely at the unit-opening photograph. Write down 5 things you observe about it. You will return to this image in the AP® Seeing Geography feature at the end of the unit.

1)

2)

3)

4)

5)

Module 44

The Origin and Influences of Urbanization

Before You Read the Module

Focus on Learning Goals

Use the Module learning goals as a way to divide or "chunk" the modules into parts. After you read a section of the module and answer the questions in this book, you should stop and ask yourself: "do I sufficiently understand the goal?"

LG 44-1: Explain how and where cities originated.

LG 44-2: Explain which processes influence patterns of urbanization.

Key Terms

Take a look at the key terms you will encounter in this module.

agricultural surplus
capitalism
city
communism
first urban revolution
redevelopment
second urban revolution
socioeconomic stratification
site
situation
streetcar
urban
urban hearth areas

While You Read the Module

Answer the questions as you read Module 44: The Origin and Influences of Urbanization.

LG 44-1: ***Explain how and where cities originated.***

1) Complete the following table. Describe how each of the following led to the origin of cities

Agricultural Surplus	
Socioeconomic Stratification	

2) Define *urban hearth areas* and identify the location of early urban hearths.

3) Describe the *site* factors that influenced how and where cities originated.

4) Describe the *situation* factors that influenced how and where cities originated.

5) Describe the two different ways in which urbanization diffused.

LG 44-2: *Explain which processes influence patterns of urbanization.*

6) Use Figure 44.8 (p. 641) to answer the following question. What is the spatial disruption of China's largest cities?

7) Describe how industrialization influenced rural-to-urban migration.

8) Describe how innovations in transportation and communication have shaped the layout of cities over time.

9) Describe how the second urban revolution reshaped cities.

10) Define *redevelopment*.

After You Read the Module

Review the Key Terms

Match each key term on the left with its definition on the right.

____ agricultural surplus	1) The agricultural and socioeconomic innovations that led to the rise of the earliest cities.
____ capitalism	2) An economic and political system in which all property is publicly owned and managed.
____ city	3) The relative location of a place about its surrounding features, or its regional position with respect to other places.
____ communism	4) A relatively large, densely populated settlement with a much larger population than rural towns and villages, cities serve as important commercial, governmental, and cultural hubs for their surrounding regions.
____ first urban revolution	5) A set of activities intended to revitalize an area that has fallen on hard times.
____ redevelopment	6) Crop yields that are sufficient to feed more people than the farmer and their family.
____ second urban revolution	7) An economic and political system in which a country's trade and industry are controlled by private owners for profit rather than owned and run by the state.
____ site	8) Relating to a city.
____ situation	9) The industrial innovations in mining and manufacturing led to increased urban growth.
____ urban	10) An absolute location of a place on Earth.

Module 45
Cities Across the World

Before You Read the Module

Focus on Learning Goals

Use the Module learning goals as a way to divide or "chunk" the modules into parts. After you read a section of the module and answer the questions in this book, you should stop and ask yourself: "do I sufficiently understand the goal?"

LG 45-1: Define contemporary cities using quantitative data.
LG 45-2: Describe the spatial outcome of urbanization in the contemporary world.
LG 45-3: Explain new urban land uses and the challenges they create.

Key Terms

Take a look at the key terms you will encounter in this module.

automobile cities
boomburb
decentralize
edge city
exurb
infill development
megacities
metacities
metropolis
metropolitan statistical area
micropolitan statistical area
sprawl
suburb
suburbanization
urban area
urbanization rate

While You Read the Module

Answer the questions as you read Module 45: Cities Across the World.

LG 45-1: ***Define contemporary cities using quantitative data.***

1) Define metropolis.

2) Explain how China and the United States define cities differently.

LG 45-2: ***Describe the spatial outcome of urbanization in the contemporary world.***

3) Define the urbanization rate.

4) Using Figure 45.6 (p. 652), identify which regions of the world have the greatest number of megacities.

5) Describe the difference between metacities and megacities.

LG 45-3: *Explain new urban land uses and the challenges they create.*

6) Define *suburbanization*.

7) Describe how suburbanization creates sprawl.

8) Describe the development of edge cities.

9) Describe boomburbs and the process of infill development.

10) Define exurbs.

After You Read the Module

Review the Key Terms

Match each key term on the left with its definition on the right.

____ automobile cities	1) A city with a regional population of over 20 million.
____ boomburb	2) A very large and densely populated city, particularly the capital or major city of a country or region.
____ decentralize	3) The building of new retail, business, or residential spaces on vacant or underused parcels in already-developed areas.
____ edge city	4) A place with more than 100,000 residents that is not a core city in a metropolitan area; a large suburb with its government.
____ exurb	5) The percentage of a nation's population living in towns and cities.
____ infill development	6) In an urban context, to move business operations from core city areas into outlying areas such as suburbs.
____ megacities	7) The movement of people from urban core areas to the surrounding outskirts of a city.
____ metacities	8) The tendency of cities to grow outward in an unchecked manner.
____ metropolis	9) A semi-rural district located beyond the suburbs that is often inhabited by well-to-do families.
____ sprawl	10) Cities whose size and shape are dictated by and almost require individual automobile ownership.

____ suburbanization

11) A concentration of business, shopping, and entertainment that developed in the suburbs, outside of a city's traditional downtown or central business district.

____ urban area

12) Any self-governing place in the United States that contains at least 5000 people.

____ urbanization rate

13) A city with a regional population of over 10 million.

Rank the following urban settlements from largest (1) to smallest (4).

____ megacities

____ urbanized area

____ megacities

____ city

Module 46
Cities and Globalization

Before You Read the Module

Focus on Learning Goals

Use the Module learning goals as a way to divide or "chunk" the modules into parts. After you read a section of the module and answer the questions in this book, you should stop and ask yourself: "do I sufficiently understand the goal?"

LG 46-1: Describe world cities and their roles in globalization.

LG 46-2: Explain the networks and linkages that drive globalization.

Key Terms

Take a look at the key terms you will encounter in this module.

gated community | world cities

While You Read the Module

Answer the questions as you read Module 46: Cities and Globalization.

LG 46-1: ***Describe world cities and their roles in globalization.***

1) Define *world cities.*

2) Describe world cities and their role in globalization.

3) Describe gated communities.

4) What spatial patterns are shown on the map in Figure 46.3 (p. 662)?

LG 46-2: ***Explain the networks and linkages that drive globalization.***

5) Explain how transportation services drive globalization.

6) Describe how communications systems and business services drive globalization.

After You Read the Module

Review the Key Terms

Match each key term on the left with its definition on the right.

____ gated community	1) A city that is a control center of the global economy, in which major decisions are made about the world's commercial networks and financial markets.
____ world city	2) The privately governed and highly secure residential area within the bounds of a city often has a fence or a gate surrounding it.

Correctly match the following networks with the examples on the right.

____ transportation services	1) media organizations that provide fast-breaking local news, international news, and information services such as global stock market reports.
____ communication systems	2) Railroads, highway networks, harbors, and international airports.
____ business services	3) stock exchanges, bond markets, and foreign currency markets.

Identify whether the following statements are true or false.

T / F	1) The world's largest cities are always considered world cities.
T / F	2) The largest number of individuals with assets of $30 million or more live in the United States, Europe, and East Asia.

T / F

3) World cities drive globalization through networks and linkages in transportation services, communications systems, and business services.

T / F

4) The world's top 25 most influential global cities are located in Europe, East Asia, and Southeast Asia.

Module 47
The Size and Distribution of Cities

Before You Read the Module

Focus on Learning Goals

Use the Module learning goals as a way to divide or "chunk" the modules into parts. After you read a section of the module and answer the questions in this book, you should stop and ask yourself: "do I sufficiently understand the goal?"

> **LG 47-1:** Describe the urban systems.
> **LG 47-2:** Explain how the rank-size rule and primate city rule help to explain the urban hierarchy.
> **LG 47-3:** Describe central place theory.
> **LG 47-4:** Describe the gravity model.

Key Terms

Take a look at the key terms you will encounter in this module.

central place
central place theory
gravity model
primate city
rank-size rule
range
threshold
urban hierarchy
urban system

While You Read the Module

Answer the questions as you read Module 47: The Size and Distribution of Cities.

LG 47-1: ***Describe the urban systems.***

1) Define urban system.

2) Describe how the state of Washington illustrates an urban system.

LG 47-2: ***Explain how the rank-size rule and primate city rule help to explain the urban hierarchy.***

3) Define urban hierarchy.

4) Explain the rank-size rule.

5) Explain primate cities.

LG 47-3: ***Identify the different parts of a map.***

6) Describe Christaller's Central Place Theory.

7) Identify the five assumptions of Christaller's Central Place Theory.

8) Explain the relationship between threshold and range.

9) Describe the relevance of the central place theory in the contemporary world.

LG 47-4: ***Describe the gravity model.***

10) Define the gravity model.

11) Describe the relevance of the gravity model in the contemporary world.

12) Describe the criticism of the gravity model.

After You Read the Module

Review the Key Terms

Match each key term on the left with its definition on the right.

____ central place	1) A city that is much larger than any other city in the country and that dominates the country's economic, political, and cultural life.
____ gravity model	2) A set of interdependent cities or urban places connected by networks.
____ primate city	3) In central place theory, the number of people required to support businesses.
____ rank-size rule	4) The idea is that the closer the two places are, the more they will influence each other.
____ range	5) A settlement that makes certain types of products and services available to consumers.
____ threshold	6) A ranking of cities, with the largest and most powerful cities at the top of the hierarchy.
____ urban hierarchy	7) In central place theory, the distance people will travel to acquire a good.
____ urban system	8) The population of a settlement is inversely proportional to its rank in the urban hierarchy.

Provide an example for each of the following.

Services with a small threshold and range	
Services with a large threshold and range	

Module 48
The Internal Structure of Cities

Before You Read the Module

Focus on Learning Goals

Use the Module learning goals as a way to divide or "chunk" the modules into parts. After you read a section of the module and answer the questions in this book, you should stop and ask yourself: "do I sufficiently understand the goal?"

> **LG 48-1:** Explain the models and theories that describe the internal structure of North American cities.
>
> **LG 48-2:** Explain the models and theories that describe the internal structure of cities in Latin America, Southeast Asia, and sub-Saharan Africa.

Key Terms

Take a look at the key terms you will encounter in this module.

bid rent theory
concentric zone model
galactic city (peripheral) model
gentrification
Giffin-Ford model
Hoyt (sector) model
multiple-nuclei model
Southeast Asian city model
sub-Saharan Africa city model

While You Read the Module

Answer the questions as you read Module 48: The Internal Structure of Cities.

LG 48-1: ***Explain the models and theories that describe the internal structure of North American cities.***

1) Explain the structure of the concentric zone model.

2) Describe the structure of the Hoyt sector model.

3) Describe the structure of the multiple-nuclei model.

4) Describe the structure of the galactic city (peripheral) model.

5) Explain how the bid rent theory can be applied to urban geography.

LG 48-2: ***Explain the models and theories that describe the internal structure of cities in Latin America, Southeast Asia, and sub-Saharan Africa.***

6) Explain the structure of the Latin American city model.

7) Describe the structure of the Southeast Asian city model.

8) Describe how the Southeast Asian model differs from the North American and Latin American city models.

9) Describe the structure of the sub-Saharan Africa city model.

10) Explain how colonization has influenced the development of Southeast Asian and sub-Saharan Africa models.

After You Read the Module

Review the Key Terms

Match each key term on the left with its definition on the right.

____ concentric zone model	1) A model of a city's internal organization showing residential districts organized around several nodes (nuclei) rather than one central business district.
____ galactic city (peripheral) model	2) A model of a city's internal organization that has three distinct CBDs: colonial, traditional, and market zones that are surrounded by ethnic neighborhoods and industry.
____ gentrification	3) A model of a city's internal organization that focuses on transportation and communication as the drivers of the city's layout.
____ Griffin-Ford model	4) A model of a city's internal organization shows a commercial spine as an extension of the CBD and is flanked by elite residents and a mall on one end, an industrial park on the other, and surrounded by the periferico.
____ Hoyt sector model	5) A model of a city's internal organization in which the central business district remains central, but multiple shopping areas, office parks, and industrial districts are scattered throughout the surrounding suburbs and linked by metropolitan expressway systems.
____ multiple-nuclei model	6) A model of a city's internal organization is organized in five concentric rings that model the arrangement of different residential zones radiating outward from a central business district.
____ Southeast Asian city model	7) The displacement of lower-income residents by higher-income residents occurs as an area or neighborhood improves.
____ sub-Saharan African city model	8) A model of a city's internal organization that is centered around a port that is surrounded by a mixed land-use zone, a Western commercial zone, and an alien commercial zone.

Module 49

Density, Land Use, and Infrastructure

Before You Read the Module

Focus on Learning Goals

Use the Module learning goals as a way to divide or "chunk" the modules into parts. After you read a section of the module and answer the questions in this book, you should stop and ask yourself: "do I sufficiently understand the goal?"

> **LG 49-1:** Explain how patterns of land use, population and housing density, and technological capabilities shape metropolitan regions.
>
> **LG 49-2:** Explain recent patterns in city infrastructure, economic base, and housing.

Key Terms

Take a look at the key terms you will encounter in this module.

fiscal squeeze
zoning regulations
perceived density

While You Read the Module

Answer the questions as you read Module 49: Density, Land Use, and Infrastructure.

LG 49-1: *Explain how patterns of land use, population, and housing density, and technological capabilities shape metropolitan regions.*

1) Describe the evolution of transportation within the United States.

2) Describe perceived density.

3) Define *zoning regulations*.

4) Describe how socioeconomic patterns developed based on access to transportation.

LG 49-2: ***Explain recent patterns in city infrastructure, economic base, and housing.***

5) Define *fiscal squeeze*.

6) Explain the impact on cities due to the loss of residential population and commercial establishments.

7) Describe the change in economic base and housing patterns.

After You Read the Module

Review the Key Terms

Match each key term on the left with its definition on the right.

____ fiscal squeeze	1) Occurs when city revenues cannot keep up with increasing demands for city services and expenditures on decaying urban infrastructure.
____ perceived density	2) Laws that dictate how land can be used.
____ zoning regulations	3) The general impression of the estimated number of people present in a given area.

Complete the chart below, identifying the residential patterns for six socioeconomic groups.

Residential Pattern	*Occupiers*	*Building Type*
Protected enclaves		
Gentrified areas		
Middle-class suburbs		
Working-class neighborhoods		
Ethnic enclaves		
Excluded ghettos		

Module 50
Urban Sustainability

Before You Read the Module

Focus on Learning Goals

Use the Module learning goals as a way to divide or "chunk" the modules into parts. After you read a section of the module and answer the questions in this book, you should stop and ask yourself: "do I sufficiently understand the goal?"

LG 50-1: Identify sustainable design initiatives and zoning practices in urban geography.
LG 50-2: Explain the positive and negative responses to urban design initiatives.

Key Terms

Take a look at the key terms you will encounter in this module.

anti-displacement tenant activists
built environment
compact design
de facto segregation
diverse housing options
greenbelt
New Urbanism
smart growth
slow-growth city
zoning

While You Read the Module

Answer the questions as you read Module 50: Urban Sustainability.

LG 50-1: ***Identify sustainable design initiatives and zoning practices in urban geography.***

1) Define *built environment.*

2) Define *smart growth.*

3) Define *new urbanism.*

4) Explain the difference between smart growth and new urbanism.

5) Explain how greenbelt development supports smart growth and New Urbanism.

6) Explain slow-growth cities.

LG 50-2: ***Explain the positive and negative responses to urban design initiatives.***

7) Describe positive responses to urban design initiatives.

8) Describe concerns for urban design initiatives.

After You Read the Module

Review the Key Terms

Match each key term on the left with its definition on the right.

____ anti-displacement tenant activist	1) A zone of grassy, forested, or agricultural land separating urban areas.
____ built environment	2) Policies that combat regional sprawl by addressing issues of population density and transportation.
____ compact design	3) An approach to city planning that focuses on fostering European-style cities of dense settlements, attractive architecture, and housing of different types and prices within walking distance to shopping, restaurants, jobs, and public transportation.
____ de facto segregation	4) A city that changes its zoning laws to decrease the rate at which the city spreads horizontally, to avoid the negative effects of sprawl.
____ diverse housing options	5) The human-made space in which people live, work, and engage in leisure activities daily.
____ greenbelt	6) The classification of land according to restrictions on its use and development.
____ new urbanism	7) Racial segregation that is not supported by law but is still apparent.
____ smart growth	8) Advocates for poor and working-class residents who are at risk of losing their affordable housing to new development.
____ slow-growth city	9) Development that grows up (in the form of taller buildings) rather than out (in the form of urban sprawl).
____ zoning	10) Policy that encourages building quality housing for people and families of all life stages and income levels in a range of prices within a neighborhood.

Module 51
Challenges of Urban Changes

Before You Read the Module

Focus on Learning Goals

Use the Module learning goals as a way to divide or "chunk" the modules into parts. After you read a section of the module and answer the questions in this book, you should stop and ask yourself: "do I sufficiently understand the goal?"

LG 51-1: Explain economic challenges that result from the movement of urban populations.
LG 51-2: Explain the social challenges that cities face.
LG 51-3: Identify responses to economic and social challenges in urban areas.
LG 51-4: Identify the consequences of urban renewal and gentrification.
LG 51-5: Explain how government fragmentation challenges efficiently in urban governance.

Key Terms

Take a look at the key terms you will encounter in this module.

affordability
below-market rate housing
blockbusting
environmental injustice
environmental justice
environmental racism
exclusionary zoning
fiscal imbalance
fiscal zoning
housing choice voucher program
inclusionary zoning
land tenure
mortgage
NIMBYs
redlining
social controls
squatter settlement
urban renewal
violent crime
white flight

While You Read the Module

Answer the questions as you read Module 51: Challenges of Urban Changes.

LG 51-1: ***Explain economic challenges that result from the movement of urban populations.***

1) Describe redlining.

2) Describe blockbusting.

3) Explain the impact of redlining and blockbusting on cities.

4) Explain how redlining and blockbusting impact housing affordability.

LG 51-2: *Explain the social challenges that cities face.*

5) Identify the three factors access to social services is dependent.

6) Explain the challenges of providing services as city populations grow.

7) Explain how segregated communities are disproportionately affected by high rates of violent crimes in the United States.

8) Describe social controls.

9) Define *environmental injustice.*

10) Explain how the water in Flint, Michigan is an example of environmental injustice.

11) Describe the development of squatter settlements within urban areas.

LG 51-3: ***Identify responses to economic and social challenges in urban areas.***

12) Describe how inclusionary zoning differed from exclusionary zoning.

13) Describe how inclusionary zones are a response to challenges in urban areas.

14) Describe how local food movements are a response to challenges in urban areas.

LG 51-4: ***Identify the consequences of urban renewal and gentrification.***

15) Define *urban renewal.*

16) Describe the consequences of urban renewal.

17) Define *gentrification.*

18) Describe the positive effects of gentrification on a community.

19) Describe the negative effects of gentrification on a community.

LG 51-5: ***Explain how government fragmentation challenges efficiency in urban governance.***

20) Define *fiscal imbalance*.

21) Define *fiscal zoning*.

22) Describe how the fragmentation of governments presents challenges in addressing urban issues.

After You Read the Module

Review the Key Terms

Match each key term on the left with its definition on the right.

____ affordability	1) Occurs when certain groups carry a larger share of environmental risks and hazards than groups who have the power to influence decisions about the environment.
____ blockbusting	2) The mass movement of white people from the city to the suburbs.
____ environmental injustice	3) The practice of identifying high-risk neighborhoods on a city map and refusing to lend money to people who want to buy property in those neighborhoods.
____ housing choice voucher	4) A federal government program to assist very-low-income families, the elderly, and the disabled with affordable, decent, safe, and sanitary housing.
____ redlining	5) An area of degraded, seemingly temporary, inadequate, and often illegal housing.
____ squatter settlement	6) A practice in which realtors persuade white homeowners in a neighborhood to sell their homes by convincing them that the neighborhood is declining due to Black families moving in.
____ white flight	7) The maximum price that a buyer can afford to pay for a house or apartment.

Review the Key Terms

Match each key term on the left with its definition on the right.

____ below market rate housing	1) Municipal and county planning ordinances that require a given share of new construction to be affordable for people with low to moderate incomes.
____ exclusionary zoning	2) The advocation for the consumption of food produced within a short distance, emphasizing the support of local farmers and promotes environmental sustainability.
____ fiscal zoning	3) Abbreviation for "not in my backyard"; term for people who try to prevent the construction of affordable housing and other types of development in their neighborhood.
____ gentrification	4) Housing that costs much less than the going rate.
____ inclusionary zoning	5) The process is where higher-income individuals move into lower-income urban areas, increasing property values and often displacing existing residents and businesses.
____ local food movement	6) Large-scale redevelopment of the built environment in downtown and older inner-city neighborhoods.
____ NIMBYs	7) The practice of using local land-use regulation to preserve and possibly enhance the local property tax base.
____ urban renewal	8) Zoning that attempts to keep low- to moderate-income people out of a neighborhood.

Module 52
Challenges of Urban Sustainability

Before You Read the Module

Focus on Learning Goals

Use the Module learning goals as a way to divide or "chunk" the modules into parts. After you read a section of the module and answer the questions in this book, you should stop and ask yourself: "do I sufficiently understand the goal?"

LG 52-1: Describe challenges to urban sustainability.

LG 52-2: Describe how cities can respond to the challenges of sustainability.

Key Terms

Take a look at the key terms you will encounter in this module.

bioremediation
brownfield
brownfield remediation
ecological footprint
Farmland Protection Policy Act (FPPA)
scattered development
urban footprint
urban heat island
urban risk divide

While You Read the Module

Answer the questions as you read Module 52: Challenges of Urban Sustainability.

LG 52-1: ***Describe the challenges to urban sustainability.***

1) Define *urban heat island.*

2) Describe the impact cities have on the natural environment.

3) Explain the urban footprint.

4) Using Figure 52.3 (p. 733), identify the continents with the most polluted cities.

5) Describe why natural disasters are challenges for cities of all sizes.

6) Using Figure 52.8 (p. 736), identify which region ranks highest and which region ranks lowest in planning for climate change.

LG 52-2: ***Describe how cities can respond to the challenges of sustainability.***

7) Describe the advantages of brownfield remediation.

8) Explain how the Farmland Protection Policy Act (FPPA) protects communities from scattered development.

After You Read the Module

Review the Key Terms

Match each key term on the left with its definition on the right.

____ bioremediation	1) The process of removing or sealing off contaminants so that a site may be used again without any health concerns.
____ brownfield	2) A mass of warm air in cities, generated by urban building materials and human activities, that sits over a city.
____ brownfield remediation	3) The idea that disasters and disaster risk become urban phenomena as the world's population becomes increasingly concentrated in large cities.
____ ecological footprint	4) Properties whose use or development may be complicated by the potential presence of hazardous substances or pollutants.
____ Farmland Protection Policy Act (FPPA)	5) The spatial extent of an urban area's impacts on the natural environment.
____ scattered development	6) Subdivisions or developments that do not border on existing settlements and that remove agricultural land from production.
____ urban footprint	7) The process of removing or sealing off contaminants so that a site may be used again without any health concerns.
____ urban heat island	8) The total amount of natural resources used and their impact on the natural environment.
____ urban risk divide	9) U.S. law that grants municipalities oversight over federally funded development projects on farmland.

Unit 6
Review Exercises

Check Your Understanding

Review the "Module Review" summaries at the end of each module in Unit 6 of your textbook. Make sure you understand all of the concepts before moving on to the next unit.

Unit 6 Multiple-Choice Question AP® Exam Practice

1. What two elements were necessary for the dramatic social change that led to the development of cities?
   (A) Agricultural surplus and innovations in transportation
   (B) Emergence of socioeconomic stratification and innovations in communication
   (C) An agricultural surplus that resulted in population growth and the emergence of socioeconomic stratification
   (D) Development of new technologies and government policies
   (E) The establishment of trade routes and new farming methods

2. Which of the following best describes the urbanization rate patterns across the globe?
   (A) African and Asian countries have the highest urbanization rates
   (B) Europe, North America, Latin America, and the Caribbean have relatively high levels of urbanization
   (C) The developing world's urban population is decreasing
   (D) Urbanization rates are evenly distributed across all continents
   (E) South Asia and sub-Saharan Africa have low urbanization rates and are expected to remain low

3. Which of the following is NOT typically a characteristic of a world city?
   (A) Cultural diversity
   (B) Headquarters of multinational corporations
   (C) Large population sizes
   (D) Advanced communications systems
   (E) Major international airport

4. Which of the following statements best describes the rank-size rule?
   (A)The largest city in a country has more than twice as many people as the second-largest city.
   (B) The distribution of cities is primarily influenced by their economic functions.
   (C) Urban hierarchies are more commonly observed in developing countries.
   (D) Cities are located based on their access to transportation networks and resources
   (E) The population of a settlement is inversely proportional to its rank in the urban hierarchy.

5. Which of the following is a characteristic of the Burgess concentric zone model?
   (A) It organizes cities in five concentric zones or rings radiating outward from the center of the city.
   (B) It focuses on transportation and communication as the drivers of the city's layout.

(C) It features a central business district with multiple shopping areas, office parks, and industrial districts linked by the ring road.
(D) It shows urban residential districts organized around several nodes rather than one primary CBD.
(E) It combines elements of Latin American culture with modern forces of globalization.

6. Which of the following factors has NOT significantly contributed to the decentralization of cities?
(A) Improvements in transportation and technology?
(B) Increased reliance on mass transit systems
(C) Advances in communication and internet technology
(D) The ability of businesses and employees to work remotely
(E) Development of robotic technology

7. Which of the following is a goal of smart growth initiatives?
(A) Promoting urban sprawl
(B) Encouraging the development of compact, walkable neighborhoods
(C) Decreasing population density
(D) Increasing dependence on automobiles
(E) Restriking mixed land use

8. What is the term used to describe the practice of realtors persuading white homeowners to sell their homes by convincing them that the neighborhood is declining due to Black families moving in?
(A) Redlining
(B) Gentrification
(C) Urban renewal
(D) Inclusionary zoning
(E) Blockbusting

9. Which of the following is NOT a challenge to urban sustainability?
(A) Suburban sprawl
(B) Need to provide clean water and air
(C) Climate change\
(D) Cities altering natural landscapes
(E) An abundance of natural resources

Unit 6 Free-Response Question AP® Exam Practice

1. Cities undergo constant transformations that affect their inhabitants differently. Respond to parts A, B, C, D, E, F, and G.
   A. Describe how housing discrimination has affected urban communities.
   B. Explain the issue of housing affordability in urban areas and its effects on residents and communities.
   C. Discuss the challenges related to accessing services in urban areas, particularly for low-income residents.
   D. Define urban sprawl.
   E. Explain how transportation systems historically shaped the development of cities, shaping socioeconomic patterns.
   F. Describe the concept of mixed-use development.
   G. Describe the role of infill development in promoting sustainable urban growth.

Notes:

Unit 7
Industrial and Economic Development Patterns and Processes

Unit Summary

The Industrial Revolution fundamentally reshaped human existence, impacting nearly every tangible item we use daily, from clothing to food and housing, due to industrialized economies. Beyond material goods, it profoundly transformed social lives and cultural values, shifting our perception of time and space away from natural cycles towards the rigid 24-hour clock. This era spurred rapid urbanization and faster transportation, accelerating the pace of life and prompting reconsiderations of traditional gender roles. While industrialization's effects are widespread, their intensity and speed vary globally, leading to disparities in living conditions and access to resources, often categorized as differences in levels of development.

Modules in Unit 7

Module 53: The Industrial Revolution
Module 54: Economic Sectors and Patterns
Module 55: Theories and Measures of Development
Module 56: Women and Economic Development
Module 57: Trade and the World Economy
Module 58: Changes as a Result of the World Economy
Module 59: Sustainable Development

What do you see?

Look closely at the unit-opening photograph. Write down 5 things you observe about it. You will return to this image in the AP® Seeing Geography feature at the end of the unit.

1)

2)

3)

4)

5)

Module 53
The Industrial Revolution

Before You Read the Module

Focus on Learning Goals

Use the Module learning goals as a way to divide or "chunk" the modules into parts. After you read a section of the module and answer the questions in this book, you should stop and ask yourself: "do I sufficiently understand the goal?"

LG 53-1: Describe the Industrial Revolution and how it influenced the diffusion of industrialization.

LG 53-2: Explain the importance of natural resource availability to industrialization.

LG 53-3: Describe the major demographic and social consequences of industrialization.

LG 53-4: Explain how industrialists' needs for resources and markets drove colonialism and imperialism.

Key Terms

Take a look at the key terms you will encounter in this module.

assembly line
capitalist class
commercial farmers
crude oil
fossil fuels
industrial revolution
international division of labor
labor productivity
labor unions
mass consumption
mass production
middle class
textile
working class
wage labor

While You Read the Module

Answer the questions as you read Module 53: The Industrial Revolution

LG 53-1: ***Describe the Industrial Revolution and how it influenced the diffusion of industrialization.***

1) Describe the preconditions that led to the Industrial Revolution beginning in Great Britain.

2) Define *labor productivity*.

3) Explain how the industrial revolution diffused.

LG 53-2: ***Explain the importance of natural resource availability to industrialization.***

4) Describe why the location of natural resources was important to the Industrial Revolution.

LG 53-3: ***Describe the major demographic and social consequences of industrialization.***

5) Explain the demographic consequence of the Industrial Revolution.

6) Describe how the Industrial Revolution led to the further development of social classes.

7) Explain how the Industrial Revolution led to the further development of divisions of labor.

LG 53-4: ***Explain how industrialists' needs for resources and markets drove colonialism and imperialism.***

8) Explain how the need for natural resources drove colonialism and imperialism.

9) Using Figure 53.10 (p. 764), identify which European countries had the largest influence in Africa.

10) Describe how imperialism led to an international division of labor.

After You Read the Module

Review the Key Terms

Match each key term on the left with its definition on the right.

____ assembly line	1) The average amount of goods or services produced per worker per unit of time.
____ capitalism class	2) The situation in which the labor forces of different countries and world regions play complementary roles in an interdependent global economy.
____ industrial revolution	3) The people in an industrial economy depend on wage labor to obtain the necessities of life.
____ international division of labor	4) The rapid transformation of the economy through the introduction of machines, new power sources, and new chemical processes in Europe and the United States between 1760 and 1820.
____ labor productivity	5) Associations of workers in particular industries were established to collectively bargain with capitalists.
____ labor unions	6) A system of manufacturing in which parts and procedures are added one step at a time through a series of workstations until a finished product is assembled.
____ mass production	7) People who own the means of production and pay the wages of workers.
____ textile	8) A socioeconomic relationship in which an employer pays a worker to complete a task, sometimes by the day or by the hour.
____ working class	9) The machine manufactures large quantities of identical products.
____ wage labor	10) A fabric or cloth is woven from the fibers of wool, cotton, or flax.

Module 54
Economic Sectors and Patterns

Before You Read the Module

Focus on Learning Goals

Use the Module learning goals as a way to divide or "chunk" the modules into parts. After you read a section of the module and answer the questions in this book, you should stop and ask yourself: "do I sufficiently understand the goal?"

LG 54-1: Explain the main economic sectors and their distinct development patterns.
LG 54-2: Identify the factors that influence the location of manufacturing.

Key Terms

Take a look at the key terms you will encounter in this module.

base industry
break-of-bulk-point
containerization
economic sectors
least-cost theory
primary sectors
secondary sectors
semiperiphery
shipping containers
tertiary sector
quaternary sector
quinary sector

While You Read the Module

Answer the questions as you read Module 54: Economic Sectors and Patterns.

LG 54-1: *Explain the main economic sectors and their distinct development patterns.*

1) Identify the five main economic sectors.

2) Describe the correlation between economic development and patterns of economic sectors.

LG 54-2: ***Identify the factors that influence the location of manufacturing.***

3) Describe the factors that influence the location of manufacturing

4) Define *break-of-bulk point.*

5) Describe how shipping containers impacted transportation.

6) Describe Alfred Weber's least cost theory regarding the relationship between raw materials, factories, and markets.

After You Read the Module

Review the Key Terms

Match each key term on the left with its definition on the right.

____ base industry	1) Countries or regions whose economies have elements of both the core and the periphery.
____ break-of-bulk-point	2) Standardized, stackable, intermodal metal boxes used to transport goods by ship, railroad, or truck.
____ containerization	3) Industries that extract natural resources from the environment.
____ economic sectors	4) A location where cargo is transferred from one mode of transportation to another.
____ least-cost theory	5) The portion of the economy where the highest-level management decisions are made in the areas of business, government, education, and science.
____ primary sector	6) The system of intermodal freight transport using shipping containers.
____ secondary sector	7) A portion of the economy is dedicated to intellectual and information services, such as scientific research and development.
____ semiperiphery	8) Industries that provide services to businesses and consumers, including all the different types of work necessary to transport and deliver goods and resources.
____ shipping containers	9) Alfred Weber's theory is that transportation costs and labor costs play a strong role in determining the location of manufacturing facilities.
____ tertiary sector	10) An industry of disproportionate economic importance and on whose existence other industries and employment sectors depend.

_____ quaternary sector

11) Groupings of industries based on what is produced and the activities of the workforce.

_____ quinary sector

12) Industries that process the raw materials extracted by primary industries, transform them into finished, usable forms.

Module 55

Theories and Measures of Development

Before You Read the Module

Focus on Learning Goals

Use the Module learning goals as a way to divide or "chunk" the modules into parts. After you read a section of the module and answer the questions in this book, you should stop and ask yourself: "do I sufficiently understand the goal?"

LG 55-1: Explain the different theories of economic development.

LG 55-2: Describe the different measures of economic and social development.

Key Terms

Take a look at the key terms you will encounter in this module.

commodity dependence
dependency theory
formal sector
human development index (HDI)
gender inequality index (GII)
gross domestic product (GDP)
GDP per capita
gross national income (GNI)
gross national product (GNP)
income distribution
informal sector
purchasing power parity (PPP)
world systems theory

While You Read the Module

Answer the questions as you read Module 55: Theories and Measures of Development.

LG 55-1: ***Explain different theories of economic development.***

1) Describe the characteristics of Rostow's Stages of Economic Growth.

2) Explain the characteristics of Wallerstein's World Systems Theory.

3) Describe dependency theory.

4) Define commodity dependence.

5) Using Figures 55.2 (p. 781) and 55.3 (p.782), answer the following question. How do the geographic patterns between commodity dependence and a country's position in Wallerstein's World Systems Theory?

LG 55-2: ***Describe the different measures of economic and social development.***

6) Define *Gross National Income* (GNI).

7) Describe the difference between gross domestic product (GDP) and GDP per capita.

8) Describe the gender inequality index as a measure of social growth.

9) Describe the human development index as a measure of social growth.

10) Describe the difference between the formal and informal sectors of an economy.

After You Read the Module

Review the Key Terms

Match each key term on the left with its definition on the right.

____ commodity dependence	1) The total income of a country's residents and businesses, including investment income, regardless of where it was earned, as well as money received from abroad such as foreign investment and development aid.
____ dependency theory	2) The part of the economy that is officially recorded with the government.
____ formal sector	3) The total value of all goods and services produced within a county over a specific period, regardless of the producer's national origin.
____ human development index (HDI)	4) Occurs when commodities account for more than 60 percent of the value of a country's total exports.
____ gender inequality index (GII)	5) Wallerstein's theory of economic development regards world history as moving through a series of socioeconomic systems, culminating in the modern world system by about the year 1900.
____ gross domestic product (GDP)	6) A country's GDP is divided by its total population.
____ GDP per capita	7) A statistical measure of human achievement that combines data on life expectancy at birth, education levels, and GNI per capita (PPP).
____ Gross National Income (GNI)	8) The part of any economy that is not officially recorded, monitored, or taxed by the government.
____ informal sector	9) The theoryis that the periphery is poor because it was economically dependent on the core in a disadvantageous relationship originally established under colonialism and imperialism.

____ world systems theory

10) A statistical measure of gender inequality that combines data on reproductive health, empowerment, and labor market participation.

Module 56

Women and Economic Development

Before You Read the Module

Focus on Learning Goals

Use the Module learning goals as a way to divide or "chunk" the modules into parts. After you read a section of the module and answer the questions in this book, you should stop and ask yourself: "do I sufficiently understand the goal?"

> **LG 56-1:** Explain how women's roles in society change as countries develop economically.
> **LG 56-2:** Explain the effects of women's increased presence in the workforce on gender parity.
> **LG 56-3:** Describe the role of microloans in advancing living standards and opportunities for women.

Key Terms

Take a look at the key terms you will encounter in this module.

gender empowerment measure (GEM)
microloan
gender parity

While You Read the Module

Answer the questions as you read Module 56: Women and Economic Development.

LG 56-1: ***Explain how women's roles in society change as countries develop economically.***

1) Explain how women's roles in society change as countries develop economically.

2) Using Figure 56.3 (p. 794) answer the following question, in which world regions do women hold the highest proportion of seats in national parliaments?

3) Define the gender empowerment measure (GEM).

LG 56-2: ***Explain the effects of women's increased presence in the workforce on gender parity.***

4) Define gender parity.

5) Using Figure 56.5 (p. 976) answer the following question, what relationship between the level of economic development and women's participation in the agricultural workforce is revealed in the map?

6) Explain the gender disparity in wages.

LG 56-3: ***Describe the role of microloans in advancing living standards and opportunities for women.***

7) Define microloans.

8) Describe how microloans advance living standards and opportunities for women.

After You Read the Module

Review the Key Terms

Match each key term on the left with its definition on the right.

____ gender empowerment measure (GEM)	1) A way of documenting progress toward gender equality using measures such as relative access to education, the average income for women versus men, and workforce participation.
____ gender parity	2) A measure of gender equality that includes the proportion of elected seats held by women in local governments, access to sexual and reproductive health care, women's freedom from violence, the percentage of women in economic decision-making positions, and women's participation in the labor force.
____ microloans	3) A very small loan to people with little income or collateral intended to help them establish or expand a small business.

Module 57

Trade and the World Economy

Before You Read the Module

Focus on Learning Goals

Use the Module learning goals as a way to divide or "chunk" the modules into parts. After you read a section of the module and answer the questions in this book, you should stop and ask yourself: "do I sufficiently understand the goal?"

LG 57-1: Explain the basis for international trade.
LG 57-2: Explain how neoliberal policies have stimulated globalization.
LG 57-3: Explain how economies have become more closely interconnected, even interdependent.

Key Terms

Take a look at the key terms you will encounter in this module.

absolute advantage
comparative advantage
competitive advantage
complementarity
customs union
debt crisis
financial market
free-trade agreement
imported substitution industrialization (ISI)
international monetary fund (IMF)
mercantilism
mercosur
neoliberalism
organization of petroleum exporting countries (OPEC)
protectionism
trade embargo
tariff
transnational corporation (TNC)
world bank
world trade organization (WTO)

While You Read the Module

Answer the questions as you read Module 57: Trade and the World Economy.

LG 57-1: ***Explain the basis of international trade.***

1) Define comparative advantage.

2) Define protectionism.

3) Explain patterns of global trade.

LG 57-2: ***Explain how neoliberal policies have stimulated globalization.***

4) Define neoliberalism.

5) Describe how the World Trade Organization (WTO) has fostered globalization.

6) Define customs unions.

7) Explain how regional organizations have fostered globalization.

LG 57-3: ***Explain how economies have become more closely interconnected, even interdependent.***

8) Explain how global financial crises affect all national economies.

9) Describe how economic development policies are influenced by forces beyond their control.

After You Read the Module

Review the Key Terms

Match each key term on the left with its definition on the right.

____ absolute advantage	1) A range of pro-market and anti-government positions on the economy, such as reducing government ownership and regulation and promoting privatization and market-based solutions.
____ comparative advantage	2) Tax on imported goods and services.
____ financial market	3) A theory of trade states that each country strives to export more than it imports to accumulate wealth.

____ mercantilism

4) Marketplace where financial instruments are traded; stock markets, bond markets, and foreign exchange markets are all financial markets.

____ neoliberalism

5) A country's ability to produce one product much more efficiently than it can produce other products within its economy.

____ protectionism

6) A country's ability to produce a good or service more efficiently than another country's.

____ tariff

7) Trade rules that restrict imports in order to protect domestic industries.

____ trade embargo

8) An official ban on trade with a specific country or of a specific good.

Review the Key Terms

Match each key term on the left with its definition on the right.

____ customs unions

1) A firm with the power to coordinate and control operations in more than one country, even if it does not own those operations.

____ free-trade agreement

2) An international organization that regulates trade among 164 member states, providing a framework for negotiating trade agreements and resolving trade disputes.

____ international monetary fund (IMF)

3) An international trade agreement among 12 oil-producing countries designed to regulate the output of oil.

____ mercosur

4) A South American customs union whose full members include Argentina, Brazil, Paraguay, Uruguay, and Venezuela.

____ organization of petroleum exporting countries (OPEC)

____ transnational corporation (TNC)

____ world bank

____ world trade organization (WTO)

5) A treaty between two or more countries that reduces tariffs and promotes foreign investment.

6) A free-trade agreement among two or more member countries, combined with a single, common external trade policy for nonmembers.

7) International organization that seeks to foster global monetary cooperation, achieve financial stability, facilitate international trade, and promote sustainable economic growth.

8) An international financial organization that provides funding and expertise to promote sustainable economic growth in developing countries.

Module 58

Changes as a Result of the World Economy

Before You Read the Module

Focus on Learning Goals

Use the Module learning goals as a way to divide or "chunk" the modules into parts. After you read a section of the module and answer the questions in this book, you should stop and ask yourself: "do I sufficiently understand the goal?"

LG 58-1: Explain how economic restructuring has affected industrialization and development.

LG 58-2: Explain how new production practices and technologies have reshaped the world economy.

Key Terms

Take a look at the key terms you will encounter in this module.

Agglomeration economies
corporate disinvestment
deindustrialization
export processing zone (EPZ)
fordism
free-trade zone (FTZ)
growth pole
high-technology industry
just-in-time manufacturing (JIT)
multiplier effect
new international division of labor
offshoring
outsourcing
post-fordism
special economic zones (SEZ)

While You Read the Module

Answer the questions as you read Module 58: Changes as a Result of the World Economy.

LG 58-1: ***Explain how economic restructuring has affected industrialization and development.***

1) Describe the difference between offshoring and outsourcing.

2) Explain how economic restructuring has led to the deindustrialization of core countries.

3) Explain the difference between special economic zone (SEZ) and export processing zones (EPZ).

4) Define free-trade zones (FTZ).

LG 58-2: ***Explain how new production practices and technologies have reshaped the world economy.***

5) Describe the emergence of an international division of labor.

6) Define just-in-time manufacturing.

7) Describe how agglomerations create the multiplier effect within economies.

8) Describe growth poles.

After You Read the Module

Review the Key Terms

Match each key term on the left with its definition on the right.

____ agglomeration economies	1) Specially designated duty-free area that provides warehousing, storage, and distribution facilities for goods intended for trade or re-export.
____ deindustrialization	2) Industrial zone with special incentives to attract foreign investment to places where imported materials undergo processing or assembly before being re-exported.
____ export processing zones (EPZ)	3) The decline, and sometimes complete disappearance, of employment in the manufacturing sector in the core's industrial centers.
____ Fordism	4) The economic and social arrangement based on the mass production of standardized goods, high labor union membership rates, stable and full-time manufacturing employment, and high factory wages that enable mass consumption.
____ free-trade zone (FTZ)	5) Occur where firms cluster spatially to take advantage of geographic concentrations of skilled labor and industry suppliers, specialized infrastructure, and ease of face-to-face contact with industry participants.

____ growth pole

6) Geographically pinpointed center of economic activity organized around a designated industry, commonly in the high-tech sector.

Review the Key Terms

Match each key term on the left with its definition on the right.

____ just-in-time manufacturing (JIT)	1) The spatial shift of manufacturing from developed countries to developing countries, including the global scaling of labor markets and industrial sites.
____ multiplier effect	2) Specific area within a country's borders where business and trade laws are different from those in the rest of the country.
____ new international division of labor	3) The creation of new businesses and jobs in other industries as the result of investment in a different industry.
____ offshoring	4) The shifts from manufacturing centers to spatially dispersed production sites, from standardized mass production to specialized batch production, and from a permanent workforce to temporary and contract workers.
____ outsourcing	5) The production of small batches of goods as needed by customer demand.
____ post-Fordism	6) The relocation of manufacturing and support services from one country to another.
____ special economic zones (SEZ)	7) The transfer of part of a firm's internal operations to a third party.

Module 59
Sustainable Development

Before You Read the Module

Focus on Learning Goals

Use the Module learning goals as a way to divide or "chunk" the modules into parts. After you read a section of the module and answer the questions in this book, you should stop and ask yourself: "do I sufficiently understand the goal?"

> **LG 59-1:** Explain how sustainable development policies can reduce the negative environmental effects of industrialization.
> **LG 59-2:** Explain how ecotourism can promote sustainable development.
> **LG 59-3:** Describe the United Nations Sustainable Development Goals and their roles in measuring development.

Key Terms

Take a look at the key terms you will encounter in this module.

carbon neutrality
carbon offsets
cogeneration
climate change
ecotourism
environmental pollution
nonpoint source pollution
point source pollution
resource depletion
sustainable development

While You Read the Module

Answer the questions as you read Module 59: Sustainable Development.

LG 59-1: *Explain how sustainable development policies can reduce the negative environmental effects of industrialization.*

1) Define sustainable development.

2) Explain the negative environmental effects of industrialization.

3) Explain how sustainable development policies can reduce the negative environmental effects of industrialization.

LG 59-2: ***Explain how ecotourism can promote sustainable development.***

4) Define ecotourism.

5) Describe the benefits of ecotourism in terms of sustainable development.

6) Describe the limits of ecotourism in terms of sustainable development.

LG 59-3: ***Describe the United Nations Sustainable Development Goals and their roles in measuring development.***

7) Describe the United Nations Sustainable Development Goals.

8) Explain how the UN's Sustainable Development Goals help to measure progress in development.

After You Read the Module

Review the Key Terms

Match each key term on the left with its definition on the right.

____ carbon neutrality

____ carbon offset

____ climate change

____ ecotourism

1) The consumption of natural resources faster than they can be replenished.

2) Achieving zero CO2 releases through a combination of emissions reduction and carbon removal.

3) Development that meets present consumption needs without compromising the ability of future generations to meet their consumption needs.

4) A long-term shift in global or regional climate patterns.

____ environmental pollution

____ resource depletion

____ sustainable development

5) Processes that remove or sequester (store) carbon from the atmosphere to make up for CO2 emissions elsewhere.

6) The contamination of the physical (air, water, earth) and biological components of the environment to the point that normal functions are negatively affected.

7) Travel to natural areas of ecological value in support of conservation efforts and socially just economic development.

Unit 7

Review Exercises

Check Your Understanding

Review the "Module Review" summaries at the end of each module in Unit 7 of your textbook. Make sure you understand all of the concepts before moving on to the next unit.

Unit 7 Multiple-Choice Question AP® Exam Practice

1. Which of the following was a key factor that provided an ideal launching pad for the Industrial Revolution in Great Britain?
   (A) A decrease in food production
   (B) A reduction in the availability of labor
   (C) Great Britain's leadership in global trade
   (D) A warmer, drier climate is unsuitable for textile production
   (E) The lack of technological innovation

2. Which of the following best defines the primary sector of the economy?
   (A) Industries that provide services to businesses and consumers
   (B) Industries that process raw materials into finished products
   (C) Industries focused on intellectual and informational services
   (D) Industries involving high-level management decisions
   (E) Industries that extract natural resources from the environment

3. Which of the following best characterizes Rostow's theory of economic development?
   (A) It emphasizes the unique historical and cultural contexts of individual countries.
   (B) It posits that all countries will progress through a linear sequence of five development stages.
   (C) It argues that development is primarily determined by external factors, such as colonialism.
   (D) It focuses on the role of commodity dependence in hindering economic growth.
   (E) It suggests that countries can skip stages of development based on their resources.

4. According to the World Systems Theory, which of the following is NOT a characteristic of core regions?
   (A) Advanced industrial technologies
   (B) Complex manufacturing system
   (C) High levels of consumption
   (D) Economics focused mostly on agriculture
   (E) They are industrial first

5. How does economic development influence women's roles in socicty?
   (A) It reinforces traditional roles, keeping women primarily in domestic spaces.
   (B) It leads to a decline in women's participation in the workforce.
   (C) It has no significant impact on women's social roles.

(D) It primarily increases women's involvement in unpaid labor.
(E) It expands women's opportunities in education, employment, and public life.

6. Which of the following best defines mercantilism?
   (A) A country's ability to produce goods more efficiently than another country
   (B) Trade rules that restrict imports to protect domestic industries
   (C) The idea is that countries should focus on their comparative advantage in trade
   (D) A theory where each country aims to export more than it imports to increase wealth
   (E) The policy of promoting free trade by reducing government-erected barriers

7. What are special economic zones (SEZ)?
   (A) A zone with high tariffs to protect local industries
   (B) A region with strict labor laws and high wages
   (C) A zone focused exclusively on agricultural production
   (D) A zone where trade laws are different from the rest of the country
   (E) An area designated for environmental conservation

8. What is the primary goal of sustainable development?
   (A) Meeting present consumption needs without compromising future generation's ability to meet their needs
   (B) Utilizing all natural resources as quickly as possible
   (C) Maximizing industrial production for current consumption
   (D) Ignoring the environmental effects of industrialization
   (E) Prioritizing economic growth over environmental conservation

Unit 7 Free-Response Question AP® Exam Practice

Respond to parts A, B, C, D, E, F, and G.

A. Define the concept of gender parity and describe how it is measured.
B. Describe two ways that economic development can alter traditional women's roles in a society.
C. Explain how the Gender Empowerment Index (GEI) attempts to provide a more comprehensive measure of gender equality than simply calculating the proportion of women in national parliaments.
D. Explain how neoliberal policies have influenced international trade and globalization.
E. Describe how the growth of transnational corporations (TNCs) has changed the nature of international trade.
F. Describe two factors that have contributed to deindustrialization in core regions like the United States.
G. Explain how special economic zones (SEZs) are used to promote industrialization in the periphery.

Notes:

Notes:

Preparing for the AP® Human Geography Examination

Overview

In early May, you will take the AP® Human Geography exam. A score of 3 or above on an AP® exam is considered a passing score. If you earn a "passing score" on the exam, you may be eligible to receive credit at a college or university. This means you will have demonstrated a level of knowledge equivalent to that of students completing an introductory human geography course at the college level. You may gain credit hours, advanced placement in a course, and possibly a reduction in college tuition.

This Strive Guide offers an opportunity to study closely for the exam. The textbook is aligned to the AP® Course and Exam Description, and this guide is, too. That means that it covers all the major topics and skills you will be expected to understand and implement on the exam. After working through this Guide, you will be very well positioned to earn a high score on the exam.

Exam Format

The AP® Human Geography Exam is divided into two sections: multiple choice-questions and free-response questions.

Multiple-Choice Section

The multiple-choice section consists of 60 questions and makes up 50% of your grade. You will have 60 minutes to complete this section. The multiple-choice questions include both individual and set-based questions. About 30-40% of them will reference a graph, image, map, table, or other kind of stimulus.

This section of the exam will assess the three big ideas in the course: Patterns and Spatial Organization, Impacts and Interactions, and Spatial Processes and Societal Change. It will also cover each of the seven units in the textbook. In this section, 8-10% of questions will cover the content in Unit 1, and 12-17% of questions will cover content in each of Units 2-7. Finally, this section will assess the major skill categories that are covered in the Skills Workshops in the textbook. The skills are weighted in the following ways: Concepts and Processes (25-36%), Spatial Relationships (16-25%), Data Analysis (13-20%), Visual Analysis (13-20%), and Scale Analysis (13-20%).

Free-Response Section

The free-response section consists of three questions and makes up 50% of your grade. Each free-response question (FRQ) is worth 7 points. You will have 75 minutes to complete this section. FRQ 1 will contain no stimuli (image, map, table, etc.), FRQ 2 will contain one stimulus, and FRQ 3 will include two stimuli. Each of the free response questions will assess at least two units of instruction. The skills, which are covered in the Skills Workshops in the textbook, are weighted in the following ways in this section: Concepts and Processes (23-29%), Spatial Relationships (33-43%), Data Analysis (10-19%), Source Analysis (10-19%), and Scale Analysis (10-14%).

Task Verbs

As noted in the textbook, there are five task verbs that frequently appear in the free response section. They are:

Compare

Define

Describe

Explain

Identify

Use the Task Verbs Skills Workshop on pages 8-9 in the textbook to refresh your memory on what to do for each of these task verbs.

Practical Tips for Exam Day

As the exam approaches, consider these tips for the day of the test (including the night before).

The night before the exam, make sure to eat a nutritious meal and get a good night's sleep. Sleep is essential for consolidating information and will help you perform on the exam.

The day of the exam, be sure to eat a nutritious breakfast and arrive at the exam on time. Bring a snack or some water if your school allows you to.

While taking the exam:

- Be sure to read each question carefully before responding. If you don't know the answer, click "Mark for Review" and come back to it at the end. You can use the question view to navigate to specific questions. There is no penalty for a wrong answer, so attempt to answer every question or make a guess.
- Just like you would be able to on a written exam, you can highlight/underline text and write down notes. You can do this by highlighting the text and choosing a highlight color or an underline. You may also write a note to yourself on that text or on your scratch paper, which will not be scored. You can also cross out answer choices that you believe are incorrect, just like you would be able to on a written exam.
- If you want to look more closely at a map or image, you can adjust the size of the onscreen panels to enlarge it. Starting in 2026, the APHG exam will use colored stimuli.
- When you've finished answering the multiple-choice questions, a question review screen will show you the questions you have marked for review as well as any questions you may not have answered yet. You can revisit these questions and change your answers until the timer expires.
- When answering the free response questions, you will enter your answers in the text editor on the right side of the screen. You can toggle between questions as needed. Be sure to answer all parts of the questions. With the new dimensional (point-by-point) scoring of the FRQs, it is helpful to label your answers A through G, corresponding to the questions. Remember: while good writing is appreciated by exam readers, spelling and grammatical mistakes will not count against your score.
- There is a timer on screen that you can choose to show or hide. If you hide the timer, it will reappear in the last 5 minutes of the exam to let you know the section is almost over.
- There will be a ten minute break in between the multiple choice and free response sections.
- Your eyes may start to feel tired from looking at the computer screen for more than 30 minutes. Try the 20-20-20 rule if this is the case: every 20 minutes, look at something 10 feet away for 20 seconds. Don't forget to blink!

After the exam, be sure to celebrate the hard work you put into this course!

Notes:

AP® Practice Exam 1

SECTION I: Multiple-Choice Questions

1. A geographer analyzing a map that shows variations in population density by country would most likely be using which type of map?
   (A) Reference map
   (B) Isoline map
   (C) Topographic map
   (D) Choropleth map
   (E) Polar projection

2. The Mercator projection is widely used for navigation despite its distortions. What is the primary disadvantage of this projection?
   (A) It distorts the shape of continents near the poles, making them appear much larger than they are in reality.
   (B) It eliminates all spatial distortions, making it the most accurate representation of Earth.
   (C) It does not allow for the accurate measurement of absolute distance.
   (D) It is only useful for mapping small-scale regions such as city districts.
   (E) It shows landmasses at their true sizes but distorts oceanic distances.

3. A geographer wants to emphasize the disproportionate impact of a disease outbreak by visually distorting geographic areas based on the number of reported cases. Which type of map would be most effective?
   (A) Dot density map
   (B) Topographic map
   (C) Reference map
   (D) Cartogram
   (E) Compass rose

4. Which of the following scenarios best demonstrates the concept of absolute location?
   (A) The United States is located between Canada and Mexico.
   (B) Chicago is a major trade hub due to its access to transportation networks.
   (C) The Empire State Building is located at 40.748817° N, 73.985428° W.
   (D) People in France and Canada both speak French, illustrating cultural diffusion.
   (E) The Mississippi River serves as a natural border between several U.S. states.

5. Environmental determinism and possibilism are two theories used to explain human–environment interactions. Which scenario best illustrates how human agency and possibilism can adjust to environment?
   (A) A desert society relies solely on subsistence farming and is unable to develop advanced industries due to the harsh climate.
   (B) A mountainous region remains sparsely populated because the terrain makes large-scale agriculture impossible.

(C) A coastal city thrives economically by developing advanced desalination technology to provide fresh water despite limited natural resources.
(D) A tropical nation's economy is primarily based on plantation agriculture, as its climate dictates what can be grown.
(E) A small island nation remains isolated from global trade due to its remote location and lack of arable land

6. When distinguishing between non-renewable and renewable resources, which statement is most accurate?
(A) Non-renewable resources are naturally replenished over a short period, making them sustainable.
(B) Renewable resources are infinite and never require management because they regenerate automatically.
(C) Non-renewable resources exist in finite quantities and cannot be quickly replaced once depleted, such as fossil fuels.
(D) Renewable resources can only be harvested from untouched, pristine ecosystems.
(E) Non-renewable resources are exclusively used for energy production, while renewable resources are used only for manufacturing.

7. How does the concept of glocalization apply to multinational corporations like McDonald's or Burger King?
(A) They standardize their menu across all locations to maintain brand consistency.
(B) They refuse to adapt their offerings to local consumer preferences.
(C) They modify products to reflect local tastes and cultural preferences while maintaining global branding.
(D) They rely only on local businesses to supply their ingredients.
(E) They operate independently without influence from regional markets.

8. What is the main reason contested boundaries often emerge?
(A) Political, cultural, or religious disagreements over land or governance.
(B) Differences in climate that lead to uninhabitable regions.
(C) The presence of natural resources that make certain areas unlivable.
(D) The inability of modern maps to accurately depict political borders.
(E) The universal agreement on national sovereignty.

9. Why are perceptual (vernacular) regions more difficult to define than formal or functional regions?
(A) They are based on subjective perceptions rather than measurable data.
(B) They have clearly defined political boundaries.
(C) They are only used by geographers to classify regions for research.
(D) They are primarily defined by economic factors.
(E) They always align with national borders.

10. In the twentieth-century, more than 6 million African Americans left rural southern states for factory and service jobs in the northern cities of the United States. The key pull factor in this Great Migration is
(A) Factory and service jobs in the urban north

(B) Segregationist policies in the South
(C) Hot summer temperatures in the South
(D) Cold winters in the north
(E) Better transportation infrastructure connecting northern and southern states

11. The average number of people per unit area of arable land is called
(A) Arithmetic density
(B) Agricultural density
(C) Crude density
(D) Population density
(E) Physiological density

12. Fashion styles spread among major cities; hashtags go viral when influencers use them and amplify the initial message. The process in which ideas leapfrog from one influential person to another or from one important place to another, skipping other persons or places, is best identified as which type of diffusion?
(A) Expansion
(B) Relocation
(C) Stimulus
(D) Contagious
(E) Hierarchical

13. Assimilation, acculturation, multiculturalism, and syncretism are best identified as
(A) Types of migration
(B) Effects of centrifugal forces
(C) Effects of diffusion
(D) Challenges to state sovereignty
(E) Challenges to sustainability in urban areas

14. Which of the following is a physical factor that might serve as a barrier to cultural diffusion?
(A) Mountain range
(B) Cultural taboo
(C) Legal code
(D) Political institution
(E) Access to technology

15. Local clay and vegetation are used as building materials in adobe homes such as the one shown in Figure 18.1 (p. 241). Adobe homes are most likely to be found in which region of the United States?
(A) Northwest
(B) Midwest
(C) New England
(D) Southeast
(E) Southwest

16. Native Americans in the United States and the Aboriginal peoples of Australia are examples of which type of cultures?
(A)Heterogeneous
(B) Indigenous
(C) Popular
(D) Urban
(E) Extinct

17. Advances in transportation such as those shown in Figure 22.5 (p. 287) have progressively reduced the time it takes to travel between places in a phenomenon best identified as
(A) Distance decay
(B) Creolization
(C) Contagious diffusion
(D) Time-space convergence
(E) Central place theory

18. Which of the following best explains a historical reason why about half of the world's population speaks a language that belongs to the Indo-European language family?
(A) English was the language of the early users of the Internet in the 1980s and 1990s.
(B) Europe's empire-building efforts coincided with rapid developments in transportation and communication technologies that enabled the global spread of Indo-European languages such as English, Spanish, and French.
(C) Significant out-migration of people from Europe during the Industrial Revolution caused languages from the Indo-European language family to diffuse through relocation migration.
(D) Languages in the Indo-European language family are relatively easy for people who speak languages in another language family to learn.
(E) Languages in the Indo-European language family are least likely to be classified as endangered languages.

19. Sense of place is likely to be strongest while visiting the location shown in Figure 23.6 (p. 299) for a person who practices which of the following religions?
(A) Sikhism
(B) Buddhism
(C) Hinduism.
(D) Christianity
(E) Islam

20. Which of the following world regions is the hearth of both Hinduism and Buddhism?
(A) South Asia
(B) East Asia
(C) Western Europe
(D) Southwest Asia
(E) North Africa

21. Which of the following best describes a stateless nation?
(A) A nation that lacks international recognition but governs itself independently
(B) A political entity where multiple ethnic groups share equal political power
(C) An ethnic group which aspires to have sovereignty but lacks an independent area they control
(D) A territory controlled by a foreign power with limited self-governance
(E) A nation that has recently gained full independence and established defined borders

22. Which of the following best describes the nation-state ideal in political geography?
(A) A state with multiple ethnic groups sharing power
(B) A region with complete economic independence
(C) A political unit where one nation aligns within a state's boundaries
(D) A territory controlled by international organizations
(E) A country containing multiple national, ethnic, and religious groups

23. Which characteristic is most commonly associated with core areas in political geography?
(A) They are typically located on national borders
(B) They serve as territorial nuclei from which countries expand
(C) They are always the most populated regions
(D) They develop after a country achieves independence
(E) They must be located in coastal regions

24. What makes Eastern Europe a shatterbelt region?
(A) Its location near major bodies of water
(B) Its abundant natural resources
(C) Its persistent fragmentation due to devolution and centrifugal forces
(D) Its unified political structure
(E) Its consistent economic growth patterns

25. Which type of boundary is typically drawn by external powers without regard for existing cultural or physical features, often resulting in conflict?
(A) Subsequent boundary
(B) Geometric boundary
(C) Consequent boundary
(D) Antecedent boundary
(E) Superimposed boundary

26. Which of the following best describes gerrymandering?
(A) The process of counting votes in federal elections
(B) A method of allocating Electoral College votes
(C) The manipulation of district boundaries for political advantage
(D) The natural evolution of voting districts over time
(E) The process of establishing new voting precincts

27. In electoral geography, what is the difference between "packing" and "cracking" as redistricting strategies?
(A) Packing slits voters across districts while cracking combines them
(B) Packing creates urban districts while cracking creates rural ones
(C) Packing concentrates opposition voters while cracking disperses them
(D) Packing affects state elections while cracking affects federal ones
(E) Packing increases voter turnout while cracking decreases it

28. Which of the following best describes supranationalism?
(A) The process of moving from a non-democracy to a democracy
(B) A collection of nation states relinquishing some sovereign rights to a larger governing body
(C) An economic partnership between neighboring countries
(D) The formation of new independent states
(E) The process of cultural integration

29. How does uneven development act as a centrifugal force in states?
(A) It promotes national unity through shared economic goals
(B) It creates social cohesion among different regions
(C) It strengthens the central government's control
(D) It creates tensions between developed core regions and less developed peripheral areas
(E) It encourages equal distribution of resources

30. Which of the following best describes a significant difference between intensive and extensive agricultural practices?
(A)Intensive agricultural practices always involve the use of much more land than hired labor.
(B) Relative to intensive agricultural practices, extensive agricultural practices require little hired labor or monetary investment to successfully raise crops and animals.
(C) Extensive agricultural practices are only practiced in commercial agriculture.
(D) Intensive agricultural practices are generally found in regions with low population density.
(E) Intensive agricultural practices rarely occur in tropical climates.

31. Which of the following is most likely to be used to characterize wheat farming as an extensive farming practice?
(A) The value of wheat varies greatly depending on whether it is produced by a subsistence farmer or a commercial grain farmer
(B) Subsistence farmers produce wheat on a much smaller scale than corporations
(C) High levels of capital are needed to finance planting and harvesting of wheat
(D) Inputs per unit of land are low
(E) Large-scale wheat farming is very lucrative

32. Which best describes the impact of the Columbian Exchange?
(A) Plants and animals were domesticated.
(B) The first integrated use of domesticated animals and plants greatly eased manual labor required by early farmers.
(C) The global spread of various plants and animals

(D) Increased food production more people available for work in factories
(E) Increased use of chemicals to facilitate food production led to soil and water pollution.

33. The seed drill reflected in Figure 36.1 (p. 514) increased agricultural productivity during which of the following?
(A) The First Agricultural Revolution
(B) The Columbian Exchange
(C) The Second Agricultural Revolution
(D) The Green Revolution
(E) The Third Agricultural Revolution

34. Which best describes the Green Revolution?
(A) Large-scale adoption of environmentally sustainable fertilizers and pesticides
(B) Domestication of plants and animals
(C) Development and diffusion of agriculture
(D) Transfer of crops and animals between Africa, the Americas, and Europe
(E) Diffusion of advances in agricultural technologies from more economically developed countries to developing countries such as India and Mexico that were primarily dependent on subsistence agriculture

35. Shifting cultivation is predominantly found in which climate type according to the map in Figure 38.1 (p. 537)?
(A) Mediterranean
(B) Tropical
(C) Desert
(D) Humid subtropical
(E) Steppe

36. Which best describes a benefit of global agribusiness for consumers in more industrialized countries?
(A) Higher wages for farmers in less industrialized countries
(B) A larger share of the income of consumers living in more industrialized countries is spent on purchasing food
(C) Growth in supranationalism
(D) Access to a wide variety of fresh fruits and vegetables throughout the year
(E) Lower wages for farmers in more industrialized countries

37. Identify the agricultural practice shown in Figure 41.15 (p. 584).
(A) Terraced farming
(B) Pastoral nomadism
(C) Irrigation
(D) Shifting cultivation
(E) Deforestation

38. Which of the following best describes the term situation in the context of urban development?
(A) The absolute location of a city, including its physical features like terrain and harbors
(B) The movement of people from rural areas to cities
(C) The economic and political system in which a country's trade and industry are controlled by provides owners for profit
(D) A set of activities intended to revitalize an area that has fallen on hard times
(E) The relative location of a city to its surrounding features, such as trade routes and population centers

39. What is a significant factor that contributed to the development of suburbs in the 20th century?
(A) Increased reliance on public transportation
(B) Decrease car ownership among the middle class
(C) The construction of edge cities
(D) Increased number of people working from home
(E) Increased car ownership allowed people to live further from the CBD

40. What two regions contain the largest about of megacities?
(A) South Asia and Europe
(B) Europe and the United States
(C) Europe and Southeast Asia
(D) South Asia and Southeast Asia
(E) South America and Europe

41. According to Christaller's central place theory, why do people travel short distances?
(A) To access specialized high-order goods and services.
(B) To visit large regional cities.
(C) To purchase expensive goods.
(D) To acquire everyday low-priced goods.
(E) To utilize international marketplaces

42. According to the Hoyt sector model, where do high-income groups typically reside?
(A) In the zone of transition closest to the CBD.
(B) In the areas between manufacturing sectors and middle-income sections.
(C) In the mixed land-use zone surrounding commercial zones.
(D) In new homes on large plots of land located along transportation lines radiating outward from the city center.
(E) In the zone of maturity that rings the CBED/market center.

43. What is the fiscal squeeze in the context of urban development?
(A)When city revenues are insufficient to meet the demands for city services and infrastructure maintenance.
(B) A period of economic growth due to increased tax revenues.
(C) A decrease in demand for city services due to suburban migration
(D) Increased federal funding for urban development projects
(E) A city's strategic plan to lower property taxes to attract businesses

44. What is the potential negative response to urban design initiatives like New Urbanism?
(A) Increase property values
(B) Reduction in traffic and crime
(C) Fewer restrictions on land use for property owners
(D) Concerns that affordable housing will decrease
(E) Decrease sprawl

45. Which of the following is NOT a factor that affects accessibility to social services in cities?
(A) The political affiliation of the city government
(B) The availability of childcare at service locations
(C) The cost to get and use these services
(D) Where the services are located
(E) Where the people who need the service live

46. What is the term used to describe the spatial extent of an urban area's impact on the environment?
(A) Urban heat island
(B) Urban footprint
(C) Ecological footprint
(D) Brownfield remediation
(E) Scattered development

47. What was the significant consequence of the increased food supplies and population growth during the Industrial Revolution?
(A) A decrease in rural-to-urban migration.
(B) Improved nutrition and reduced mortality rates.
(C) A decline in the population of Great Britain.
(D) An increase in problems related to food distribution.
(E) A shift towards more labor-intensive agricultural practices.

48. What is a characteristic of a country with a workforce heavily involved in the primary sector?
(A) It is considered a highly developed economy.
(B) It has a large number of workers in service industries.
(C) It is likely to have a strong manufacturing base.
(D) It is considered to be among the least developed economies.
(E) It has a high proportion of workers in technology and innovation.

49. Which of the following factors is NOT typically a major influence on manufacturing location?
(A) Distance from the equator.
(B) Availability of energy resources.
(C) Proximity to raw materials.
(D) Labor costs and availability.
(E) Access to consumer markets.

50. What is a key criticism of the dependency theory?
(A) It pays too little attention to social and cultural variations within the periphery and the core.

(B) It overemphasizes the role of core countries in the underdevelopment of peripheral countries.
(C) It does not adequately address the issue of commodity dependence.
(D) It fails to recognize the importance of trade relations between countries.
(E) It does not account for the influence of multinational corporations on developing economies.

51. What is a key characteristic of contemporary global trade?
(A) Most trade occurs between countries rather than within transnational corporations.
(B) Protectionism has increased the flow of exports and imports
(C) The growth of world GDP closely matches the increase in world merchandise trade volume.
(D) Developing economies' annual growth rates in trade exceed those of developed countries.
(E) International trade has decreased under globalization.

52. What is the primary function of the World Trade Organization (WTO)?
(A) To provide loans to countries that cannot get loans from private banks.
(B) To promote sustainable economic growth in developing countries.
(C) To foster global monetary cooperation and achieve financial stability.
(D) To regulate trade among member states, provide a framework for negotiating agreements, and resolve disputes.
(E) To control the global oil supply and oil prices.

53. Which of the following measures specifically combines data on reproductive health, empowerment, and labor market participation to assess development?
(A) Gross National Product (GNP)
(B) Gross Domestic Product (GDP)
(C) Purchasing Power Parity (PPP)
(D) Human Development Index (HDI)
(E) Gender Inequality Index (GII)

54. Figure 58.8 (p. 821) illustrates Silicon Valley as an agglomeration economy. Which of the following is a key characteristic that supports this classification, as shown on the map?
(A) A high concentration of manufacturing facilities
(B) Proximity to international airports and universities
(C) A dominant reliance on primary sector industries
(D) Low property values and taxes
(E) A significant presence of heavy industry

55. The "hollowed-out" factory in Detroit, Michigan in Figure 58.2 (p. 816) is an example of a consequence of which economic process in core regions?
(A) Agglomeration economies
(B) Outsourcing
(C) Deindustrialization
(D) Fordism
(E) Post-Fordism

56. Using Figure 27.7 (p. 367), which of the following regions is identified as a primary shatterbelt?
(A) North America
(B) East Central Europe
(C) South America
(D) Australia
(E) Western Europe

57. The boundary between Rwanda and Burundi, which was imposed by Belgium during the colonial period without regard for existing cultural boundaries, is an example of a(n):
(A) Consequent boundary
(B) Antecedent boundary
(C) Superimposed boundary
(D) Relic boundary
(E) Geometric boundary

58. The spatial distribution of several supranational political organizations. Which of the following statements accurately describes the European Union (EU) based on Figure 31.10 (p. 432)?
(A) It is primarily composed of countries in East Asia.
(B) It includes countries in North America and Europe.
(C) It is a geographically contiguous bloc of Western and Eastern European countries.
(D) Its membership extends across Europe, Africa, and parts of Asia.
(E) It is an organization primarily focused on military alliances in the Southern Hemisphere.

59. The secession of South Sudan from Sudan, leading to the creation of the world's newest nation-state, is a direct outcome of:
(A) An international trade agreement.
(B) A successful military alliance with a core state.
(C) Devolutionary forces driven by ethnic and religious differences.
(D) The adoption of a unitary system of government.
(E) An effort to expand a supranational organization.

60. Using Figure 45.4 Urban percentage of population, 2022 (p. 651), which continent or region generally has the lowest urban percentage of population?
(A) Europe
(B) North America
(C) Latin America and the Caribbean
(D) Africa and Asia
(E) Australia

SECTION II: Free-Response Questions

1. Respond to parts A, B, C, D, E, F, and G.

   A. Define rates of natural increase.
   B. Identify and explain one factor that contributed to high population growth.
   C. Describe one pull factor that would encourage people to migrate from less developed to more developed countries.
   D. Explain the degree to which population growth patterns vary between peripheral countries to core countries.
   E. Describe ONE likely economic effect when countries have negative population growth.
   F. Describe ONE likely effect of rural-to-urban migration on the economy of a developing country.
   G. Explain ONE way the total fertility rate changes as a country shifts from an agricultural economy to a manufacturing-based economy.

A.

B.

C.

D.

E.

F.

G.

Table 1: Maritime Zones Defined by the United Nations Convention on the Law of the Sea (Table 28.1, p. 390)

Map 1: UNCLOS zones for the United States (Figure 28.14, p. 390)

2. The United Nations Convention on the Law of the Sea (UNCLOS) is an international agreement that outlines how coastal states can claim, use, and control the sea and its resources based on their proximity to the coastline.

Respond to parts A, B, C, D, E, F, and G.

A. Identify one U.S. territory in the Pacific Ocean that contributes to the extension of the United States' maritime jurisdiction shown in Map 1.

B. Using the data in Table 1, describe the benefits to coastal states of having an Exclusive Economic Zone (EEZ).

C. Explain the degree to which coastal states have greater geopolitical and economic power compared to landlocked states, using the maritime zones defined by UNCLOS.

D. Describe how the median line principle is used to resolve maritime boundary disputes between coastal states with overlapping claims.

E. Describe how patterns of cultural diffusion and syncretism have shaped the cultural landscape of a U.S. maritime territory such as Hawaii or Puerto Rico.

F. Explain how the gravity model can be used to analyze trade interactions between coastal port cities.

G. Explain a limitation of the United Nations Convention of the Law of the Sea (UNCLOS) in resolving conflicts between states in the Spratly Islands.

A.

B.

C.

D.

E.

F.

G.

Image 1: Tea plantation in central Java (Figure 33.5, p. 477)

Image 2: Forest clearing for soybeans (Figure 41.5, p. 577)

3. For most of human history, obtaining food is the main way in which humans have interacted with the physical environment. Tea production is shown in Image 1, and the clearing of land to grow soybeans in Brazil is shown in Image 2.

Respond to parts A, B, C, D, E, F, and G.

A. Define climate.
B. Identify the climate type in which plantation agriculture is typically practiced.
C. Explain how food preferences may be influenced by the physical environment.
D. Explain why the cultivation of cash crops is typically associated with monoculture.
E. Describe how the historical process of colonialism influenced the development of a global system of agriculture.
F. Explain how shifting cultivation can alter the landscape.
G. Explain the degree to which plantation agriculture can be considered an intensive farming practice.

A.

B.

C.

D.

E.

F.

G.

Notes:

Notes:

Notes:

AP® Practice Exam 2

SECTION I: Multiple-Choice Questions

1. Which of the following scenarios best illustrates the concept of relative distance?
   (A) The Grand Canyon is located at 36.1069° N, 112.1129° W.
   (B) The trip from New York City to Boston is shorter by train than by car due to traffic congestion.
   (C) A map showing the highest elevations in the Rocky Mountains uses isolines.
   (D) The scale of a map reads 1:100,000, meaning one unit on the map represents 100,000 units in real life.
   (E) The state of Texas appears much larger on a Mercator projection than on a Robinson projection

2. Which of the following best illustrates how Geographic Information Systems (GIS) contribute to spatial analysis?
   (A) A smartphone navigation app calculating the shortest route to a destination
   (B) A census report listing the total population of a country
   (C) A series of images taken from a satellite showing changes in a river's course over time
   (D) A layered digital map displaying land use, population density, and transportation networks to determine the best location for a new school
   (E) A political map displaying the boundaries of all U.S. states

3. How does the Global Positioning System (GPS) differ from Geographic Information Systems (GIS)?
   (A) GPS provides spatial data for analysis, while GIS interprets and organizes data into layers for decision-making.
   (B) GIS provides location data, whereas GPS is used primarily for predicting weather patterns.
   (C) GPS is a software-based system, while GIS relies solely on fieldwork observations.
   (D) GIS is a network of satellites, while GPS is a system for mapping political boundaries.
   (E) GPS relies on satellite imagery, while GIS functions only through aerial photography.

4. Japan's "triple disaster" (earthquake, tsunami, and a flooded nuclear power plant) required multifaceted disaster preparedness. Which of the following best illustrates how human environmental perceptions led to improved planning?
   (A) Relying solely on traditional building techniques without modern early warning systems.
   (B) Ignoring the threat of natural hazards and focusing on economic growth exclusively.
   (C) Implementing uniform planning measures that treat all hazards as having the same cause and effect.
   (D) Dismissing natural hazard risks as inevitable, thereby avoiding costly preparedness investments.
   (E) Developing innovative early-warning systems, community training, and tailored infrastructure upgrades to address the distinct challenges of earthquakes, tsunamis, and nuclear safety.

5. Increased greenhouse gas emissions have complex effects on the Earth's climatWhich of the following best describes one significant impact of these emissions?
   (A) They cause a steady decline in global temperatures, resulting in longer, harsher winters.
   (B) They intensify the greenhouse effect, leading to global warming, climate change, and associated

extreme weather events.
(C) They primarily affect localized weather patterns without contributing to long-term global changes.
(D) They increase the natural replenishment rates of renewable resources by stimulating plant growth.
(E) They have negligible effects compared to other natural climatic processes, such as volcanic eruptions.

6. Why might a global-scale analysis be misleading when examining religious distributions?
(A) It assumes that all religious groups are equally distributed across countries.
(B) It fails to account for regional and local variations within countries.
(C) It only focuses on major world religions, ignoring smaller belief systems.
(D) It does not use maps to represent spatial patterns.
(E) It does not consider historical influences on religious expansion.

7. Which of the following best demonstrates the concept of a border zone?
(A) The physical border between the U.S. and Canada, which is strictly monitored.
(B) The transition area between the United States and Mexico, where cultural traits blend.
(C) The demilitarized zone (DMZ) between North and South Korea.
(D) The city of Berlin, divided during the Cold War.
(E) The Great Wall of China, built to define imperial borders.

8. Why do mental maps vary from person to person, even when describing the same region?
(A) Mental maps are formed based on personal experiences and perceptions.
(B) Only trained geographers can develop accurate mental maps.
(C) Mental maps are based strictly on official political boundaries.
(D) They are determined solely by geographic location, not by human interpretation.
(E) They do not change over time.

9. The move of millions of Polish youths to western European countries in the two decades after Poland joined the European Union illustrates the concept of
(A) Rural-to-urban migration
(B) Internal migration
(C) Brain drain
(D) Guest workers
(E) Social mobility

10. Which of the following is a characteristic of a country in the "Age of degenerative and human-made diseases" in its epidemiological transition?
(A) Short life expectancy
(B) High infant mortality rates
(C) Diseases such as cancer, heart disease, and stroke as major causes of death
(D) Poor sanitary conditions
(E) High risk of dying from infectious diseases

11. A cultural anthropologist from the United Kingdom is studying a remote village in Papua New Guinea and encounters a ritual that seems unusual by Western standards. Instead of immediately dismissing or criticizing the practice, she spends time understanding its significance within the community's belief system and social structurThis attitude toward cultural difference could best be identified as
(A) Ethnocentrism
(B) Cultural relativism
(C) Possibilism
(D) Supranationalism
(E) Environmental determinism

12. Of the following, which would be considered part of the cultural landscape?
(A) A river
(B) A sign in South Africa that is printed in multiple languages such as Afrikaans, Xhosa, and English
(C) A desert
(D) Wildlife
(E) Mountains

13. Which of the following is most evident in Figure 20.1 (p. 266)?
(A) Food preferences
(B) Postmodern architecture
(C) Placemaking
(D) Religion
(E) Agricultural land-use

14. Using Figure 20.4 (p. 269), the Kurdish language is most likely to function in Armenia as which of the following?
(A) A centripetal force at the national scale
(B) A centrifugal force at the national scale
(C) A centripetal force at the local scale
(D) Both a centrifugal and centripetal force at the national scale
(E) Both a centrifugal and centripetal force at the local scale

15. The common practice of many countries memorializing the past through monuments that portray it as a woman and signs using the female symbol in connection with children demonstrate the significance in the cultural landscape of
(A) Religion
(B) Language
(C) Land use
(D) Gender roles
(E) Traditional architecture

16. At many McDonald's outlets outside the United States, rice-based dishes are served, including kaprao fish rice in Thailand and black sticky rice with young coconut pie in ThailanCultural diffusion in which a trait is modified to adapt to local culture is best identified as what type of diffusion?

(A) Relocation
(B) Hierarchical
(C) Stimulus
(D) Contagious
(E) Expansion

17. The expansion of a highway network into the Amazonian rainforest is most likely to lead to
(A) A significant increase in the number of people who speak the indigenous languages spoken by people living in the area in which the highway network expands
(B) A significant decrease in the number of people who speak the indigenous languages spoken by people living in the area in which the highway network expands
(C) An increase in the number of people who practice Catholicism
(D) An increase in the number of people who practice animism
(E) An increase in the number of people who speak Portuguese

18. Using Figure 23.1 (p. 294), which of the following best explains a limitation of the map in understanding the number of people who speak a language belonging to the Sino-Tibetan family?
(A) The map does not have the year in which the data was collected, making it difficult to judge its accuracy.
(B) The map does not show where specific languages are spoken, making it difficult to understand how many people speak languages that are in the Sino-Tibetan family.
(C) The map does not show the degree to which any particular language is endangered, making it difficult to predict how many people in the future will speak a language belonging to the Sino-Tibetan family.
(D) The map does not show the population of particular places, masking the number of speakers of a language family that is clustered such as the Sino-Tibetan language family as compared to one with a wider distribution, such as the Afro-Asiatic family.
(E) The map shows the distribution of major linguistic families without showing whether someone speaks a language as a native or non-native speaker, which can be misleading given the number of non-native speakers of languages belonging to the Sino-Tibetan family.

19. Which of the following pairs consists only of religions that are clustered near their hearths and have generally spread mostly through relocation diffusion?
(A) Islam and Hinduism
(B) Buddhism and Christianity
(C) Hinduism and Judaism
(D) Christianity and Islam
(E) Judaism and Christianity

20. The Korean language developed distinct dialects in North Korea and South Korea following the division of the Korean peninsula after the end of World War II. The dialect spoken in North Korea has been influenced by allies such as the Soviet Union, while cultural exchange with allies such as the United States and United Kingdom has helped shape the dialect spoken in South KoreThe development of different dialects of Korean reflects the geographic process of
(A) Assimilation
(B) Creolization
(C) Expansion diffusion
(D) Time-space compression

(E) Cultural divergence

21. Which of the following accurately describes an autonomous region?
    (A) A fully independent sovereign state
    (B) A territory with no political boundaries
    (C) A subdivision within a state with some self-governing authority
    (D) A nation with complete control over foreign policy
    (E) A region within a state that has complete independence.

22. How does ethnonationalism function as a devolutionary force?
    (A) It promotes centralized government control
    (B) It strengthens international alliances
    (C) It encourages ethnic groups to seek political autonomy
    (D) It reduces cultural differences between regions
    (E) It eliminates regional economic disparities

23. How does the Strait of Hormuz function as a global choke point?
    (A) It serves as a cultural boundary between nations
    (B) It acts as a political border between countries that serves as a buffer zone
    (C) It provides unlimited access to shipping routes
    (D) It restricts maritime traffic and controls access to vital shipping routes
    (E) It serves as a major tourist destination

24. Which statement best explains the concept of a borderland?
    (A) A region completely isolated from neighboring cultures
    (B) A territory exclusively controlled by one nation
    (C) A region where cultures blend across both sides of an international boundary
    (D) An area with strict border control policies
    (E) A territory with no cultural influences from neighboring regions

25. Which type of boundary describes a former international border that no longer serves its original function?
    (A) Relic boundary
    (B) Demarcated boundary
    (C) Delimited boundary
    (D) Subsequent boundary
    (E) Precedent boundary

26. Why might a cartogram be more effective than a traditional map for displaying presidential election results?
    (A) It shows geographic features more clearly
    (B) It represents population distribution more accurately
    (C) It highlights state boundaries better
    (D) It emphasizes rural voting patterns
    (E) It displays election data in real-time

27. What is the primary purpose of regional trading blocs?
   (A) To promote military cooperation between nations
   (B) To establish common cultural practices
   (C) To reduce or eliminate trade barriers between member states
   (D) To enforce international law
   (E) To standardize currency among member nations

28. Which best explains how nationalism functions as a centripetal force in multi-ethnic countries?
   (A) It promotes separation between different ethnic groups
   (B) It encourages regional autonomy
   (C) It creates a shared identity and common goals despite ethnic differences
   (D) It weakens central government authority
   (E) It eliminates cultural differences between groups

29. Which is a key reason global per-person protein availability has outpaced population growth in recent decades?
   (A) The growth in commercial aquaculture
   (B) Genetically modified crops
   (C) Organic farming techniques
   (D) The growth in production of value-added specialty crops
   (E) Extensive use of chemical fertilizers and pesticides

30. Which best explains how empowering women in subsistence societies is likely to impact food security on a local level?
   (A) Food security at the local level would be likely to decrease because empowering women would force men to migrate to cities to look for work.
   (B) Food security at the local level would be likely to decrease because empowering women is likely to lead to more women working in the secondary and tertiary economic sectors.
   (C) Food security at the local level would be likely to increase because empowering women is likely to produce higher crop yields.
   (D) Empowering women would likely have no impact on food security at the local level because they must own their own land in order to increase crop yields, and women are prohibited from owning land in all subsistence societies.
   (E) Food security at the local level would be likely to increase because empowering women would enable the national government to import greater amounts of food.

31. Which of the following best describes spatial patterns in the distribution of Mediterranean climate?
   (A) It is found only in East Asia and Southeast Asi
   (B) It is clustered on the entire east coast of both North and South America.
   (C) It is dispersed around the world on the eastern coasts of all continents except Antarctica.
   (D) It is clustered mostly on coastal North Africa and southern Europe but is also found in small areas of central and Southern California, central Chile, and portions of southern Australia.
   (E) It is clustered within 23 ½ degrees north and south of the Equator.

32. Which best explains how the long-lot survey system affected resource distribution?
   (A) It maximized profit by consolidating resources among wealthy landowners.
   (B) It equitably distributed access to scarce resources among the maximum number of farmers.
   (C) It minimized the number of farmers who can access agricultural land.
   (D) It prioritized large-scale commercial farming over small-scale agricultural operations.
   (E) It promoted government ownership of land in order to increase agricultural efficiency.

33. Which best describes the impact of the Fertile Crescent on modern agriculture and patterns of consumption?
   (A) Cereal grains such as wheat, barley, rye, and oats domesticated in the Fertile Crescent continue to play a key role in modern diets.
   (B) Tropical fruit was first cultivated in this region and led to the rise of plantation agriculture.
   (C) Agriculture in the Fertile Crescent had a minimal global impact due to its remote location.
   (D) The Fertile Crescent was only focused on livestock domestication and contributed to the incorporation of meat as a major source of calories in the diet of those living in Asia and Africa.
   (E) Its position along the Mediterranean coast was primarily important for fishing rather than for domestication of crops and animals.

34. An increase in agricultural productivity led to a shift from primarily subsistence farming to the production of cash crops during which of the following?
   (A) The First Agricultural Revolution
   (B) The Columbian Exchange
   (C) The Second Agricultural Revolution
   (D) The Green Revolution
   (E) The Third Agricultural Revolution

35. Which of the following explains how the demand for and price of land decrease as its distance from the central business district increases?
   (A) Distance decay
   (B) Malthusian theory
   (C) The demographic transition model
   (D) Commodity dependence
   (E) Bid-rent theory

36. Which best describes how Africa's Sahel region, which marks the transition from the Sahara Desert to more moist conditions to the south, is most likely to be impacted by global climate change?
   (A) Farmers in the region are likely to form cooperatives to take advantage of more favorable agricultural conditions.
   (B) The Sahel is likely to experience regional food shortages.
   (C) Fertility rates in all countries located in the Sahel are likely to decrease significantly.
   (D) Agricultural productivity in the region is likely to increas
   (E) Smallholder farmers in the Sahel are likely to find the most productive farmland more affordable.

37. Which best describes an economic benefit to farmers who participate in community-supported agriculture?
   (A) Farmers are able to provide food to people who live in urban areas in which suburbanization threatens the loss of farmland.
   (B) Laws in most countries allow farmers to sell produce to people who participate in community-supported agriculture at much higher prices than they could if they sold their products at a farmers' market.
   (C) If farmers are unable to provide produce to members due to local conditions such as weather extremes or even global conditions such as political conflict, members are typically not reimbursed for their subscription costs.
   (D) Farmers who participate in community-supported agriculture are required to use synthetic fertilizers and pesticides, which are much less expensive than methods used in organic farming.
   (E) Farmers who participate in community-supported agriculture only sell value-added specialty crops.

38. What is the term used to describe the building of new retail, business, or residential spaces on vacant or underused parcels of land in already developed?
   (A) Decentralization
   (B) Suburbanization
   (C) Sprawl
   (D) Exurbs
   (E) Infill development

39. What is a megacity
   (A) A city with more than 5 million residents
   (B) A city that is a world economic center
   (C) A city with more than 20 million residents
   (D) A city with more than 10 million residents
   (E) A city with a high rate of industrialization

40. The rise of gated communities in world cities is associated with
   (A) Increased emphasis on environmental quality
   (B) Decreasing income inequality
   (C) A desire to increase political participation in urban areas
   (D) The needs and tastes of the global cosmopolitan class
   (E) The expansion of public service in suburban areas

41. Which model best describes a city with a traditional CBD, a colonial CBD, and a market zone?
   (A) Griffin-Ford Model
   (B) Galactic City Model
   (C) Hoyt Sector Model
   (D) Sub-Saharan African City Model
   (E) Southeast Asian City Model

42. Which of the following is NOT a characteristic of gentrified areas?

(A) Attraction of higher-income residents
(B) Often located in formally deteriorated areas of city centers
(C) Displacement of lower-income residents
(D) Increase in affordable housing options
(E) Renovation of older family homes

43. What is the primary focus of New Urbanism?
(A) Promoting suburban development
(B) Fostering European-style cities with dense settlements and diverse housing
(C) Expanding agricultural land within urban areas
(D) Decreasing walkability in urban areas
(E) Increasing the use of large plots of land for single-family homes

44. What is the term used to describe the large-scale redevelopment of the built environment in downtown and older inner-city neighborhoods?
(A) Gentrification
(B) Inclusionary zoning
(C) Urban renewal
(D) Exclusionary zoning
(E) Blockbusting

45. What is the process of removing or sealing off contaminants in a brownfield site so it can be used again without health concerns?
(A) Bioremediation
(B) Urban remediation
(C) Ecological restoration
(D) Brownfield remediation
(E) Industrial cleanup

46. How did the Industrial Revolution affect urbanization?
(A) It led to a decrease in the proportion of the population living in cities.
(B) It had no impact on the growth of cities.
(C) It caused a transformation of small towns into urban industrial centers.
(D) It resulted in less crowded and clear urban living conditions.
(E) It slowed down the process of urbanization.

47. What is the semiperiphery in the global economy?
(A) The most economically developed countries.
(B) Countries that primarily provide raw materials.
(C) An intermediary category with elements of both core and periphery.
(D) Countries with minimal industrial activity.
(E) The sector of the economy focused on innovation and research.

48. Which of the following is a limitation of the gender inequality index (GII)
(A) It does not capture the quality of jobs for women versus men.

(B) It does not consider women's participation in national parliamentary representations.
(C) It fails to include data on reproductive health and labor market participation.
(D) It over-emphasizes unpaid labor in caregiving and housekeeping.
(E) It does not account for disparities in education levels between females and males.

49. What has been a significant outcome of microloan programs in developing countries?
(A) A decrease in women's economic independenc
(B) The establishment of profitable small businesses and improved living standards for many women.
(C) A shift towards women primarily working in large industrial sectors.
(D) And increase in women's reliance on unpaid domestic labor.
(E) Limited impact on poverty reduction or women's empowerment.

50. How did the Latin American debt crisis of the 1980s affect the region's economic policies?
(A) It led to increased import substitution industrialization (ISI)
(B) It had no significant impact on the region's development strategy.
(C) It resulted in decreased borrowing and reduced debt.
(D) It strengthened protectionist trade policies in the region.
(E) It forced governments to adopt neoliberal development policies.

51. How has the shift in manufacturing from core to periphery/semi-periphery countries affected the global economy?
(A) It has decreased the total volume of global trade.
(B) It has resulted in the deindustrialization of core regions.
(C) It has had minimal impact on the international division of labor.
(D) It has caused a uniform increase in wages and labor standards worldwide.
(E) It has strengthened Fordism and the dominant production model.

52. Which of the following best describes the economic activity shown in Figure 54.1 (p. 768)?
(A) Secondary sector manufacturing
(B) Tertiary sector services
(C) Quaternary sector research
(D) Primary sector extraction
(E) Quinary sector management

53. What is a key reason that companies might choose to locate manufacturing facilities near a break-of-bulk point?
(A) To access a highly skilled labor force
(B) To minimize transportation costs
(C) To be closer to end consumers
(D) To utilize renewable energy sources
(E) To avoid unionized labor

54. Using Figure 55.3 Commodity dependence (p. 782), and the World Economic Systems Theory, which statement is most accurate regarding the relationship between commodity dependence and the world systems model?
(A) Core countries are generally commodity-dependent.
(B) Semiperipheral countries show high levels of commodity dependenc
(C) Peripheral countries tend to exhibit high commodity dependenc
(D) Commodity dependence is unrelated to a country's position in the world systems model.
(E) Only countries with no data are part of the core.

55. According to Weber's least-cost theory, if the final product's weight is significantly greater than the raw materials' weight, where would manufacturing most likely be located to minimize transportation costs?
(A) Near the raw materials source 1
(B) Near the raw materials source 2
(C) At a break-of-bulk point
(D) Closer to the market
(E) In the semiperiphery

56. The Strait of Hormuz is considered a critical maritime choke point primarily because:
(A) It is the only sea route between the Red Sea and the Mediterranean Se
(B) It is protected by international law for centuries, ensuring safe passage for all ships.
(C) It serves as the shortest sea route from the Persian Gulf to Asian markets for oil exporters.
(D) Its importance declined after the opening of the Panama Canal.
(E) It is a narrow passage that restricts traffic to land-based regions.

57. The ethnic cleansing campaign against the Rohingya people in Myanmar highlights a significant challenge to state sovereignty stemming from:
(A) International terrorism efforts.
(B) The rise of supranational organizations.
(C) Internal devolutionary forces driven by ethnic conflict.
(D) Cross-border economic disputes.
(E) Global climate change impacts.

58. The Russian annexation of the Crimean Peninsula in 2014, and subsequent actions in southeastern Ukraine in 2022 are primary examples of which devolutionary force?
(A) Domestic terrorism
(B) Subnational terrorism
(C) Ethnic cleansing
(D) Irredentism
(E) State terrorism

59. Using Figure 45.1 Metropolitan and Micropolitan Statistical Areas of the United States (p. 648), which of the following states has the largest contiguous metropolitan statistical area?

(A) California
(B) Texas
(C) New York
(D) Illinois
(E) Florida

60. Using Figure 45.6 Map of the world's largest cities (p. 652), which region on the map has the greatest number of cities classified as "Over 30 million" or "20-29.9 million" in population agglomerations for 2024?
(A) North America
(B) South America
(C) Europe
(D) Asia
(E) Africa

SECTION II: Free-Response Questions

1. Many countries in the world, from Brazil to Vietnam, produce coffee.

Respond to parts A, B, C, D, E, F, and G.

A. Identify the climate conditions necessary for coffee production.
B. Describe an economic impact of coffee farming on producing countries.
C. Describe the impact of coffee farming on food production in producing countries.
D. Identify one way increased coffee consumption worldwide affects its production.
E. Explain one effect of increased fair trade production in coffee producing countries.
F. Explain how tariffs imposed on coffee may affect its consumption in a coffee-importing country.
G. Explain one way coffee consumption has affected urban landscape in many developed and developing countries.

A.

B.

C.

D.

E.

F.

G.

Image 1: Urban percentage of population (Figure 45.5 p. 651)

2. Densely populated settlements with much larger populations than rural towns and villages are a relatively new phenomenon. Several geographic processes drive urbanization, which is a major trend in the world today.

Respond to parts A, B, C, D, E, F, and G.

A. Identify the scale of analysis of the map.

B. Using the map, describe spatial patterns in urbanization in South Asia as compared to North America.

C. Explain how internal migration during contemporary times has influenced the size of the urban population in developing countries.

D. Describe how innovations in mining and manufacturing influenced urbanization in Western Europe beginning in the late eighteenth century.

E. Explain how advances in transportation facilitated the formation of edge cities in American metropolitan areas.

F. Explain a limitation of the map in showing the distribution of population living in urban areas at the local scale.

G. Explain the degree to which the percentage of the population living in urban areas within a particular country is related to the sectoral structure of its economy.

A.

B.

C.

D.

E.

F.

G.

Image 1: Montmartre, Paris (Figure 20.1 p. 266)

Image 2: Ecotourism on Chumbe Island, Tanzania (Figure 59.9, p. 830)

3. The images reflect the practice of tourism, which can have impacts at a variety of scales. Image 1 shows a café in a neighborhood popular with tourists in Paris, FrancImage 2 shows lodging made of sustainably harvested local materials for visitors to Chumbe Island Coral Park in Tanzania, which is on the eastern coast of Africa.

Respond to parts A, B, C, D, E, F, and G.

A. Identify the economic sector into which jobs associated with tourism are classified.
B. Describe the concept of ecotourism.
C. Explain an economic reason city planners might use placemaking to attract tourists in a location such as the Montmartre neighborhood in Paris, France.
D. Explain how tourism can impact infrastructure on a local scale.
E. Explain how ecotourism can help protect the physical environment in developing countries.
F. Describe one potential economic impact of ecotourism on the local population.
G. Explain a limitation of ecotourism when considering its impact on environmental sustainability at the global scale.

A.

B.

C.

D.

E.

F.

G.

Notes:

Notes:

Notes:

Notes: